This book is dedicated to my father and mentor,
Jay S. Riley.

"Thank you Dad!"

A special thanks to my friend and mentor, Donald Ray Warren and our Heavenly Father for giving me the strength and knowledge to do what I do. Thank You!

Warning:

This book is intended for reference only. Read the manufacturer's device manual and the facility procedures before using this information. The manufacturer or the facility may have different guidelines on how the device should be setup and calibrated. The publisher and author have made every effort to ensure that all information in this book is accurate. The publisher and author shall not be held responsible for any errors in content or for the interpretation or application of the material in this book.

Instrument Installation:

Before mounting an instrument first determine what type of service the instrument is going to be reading, liquid or gas. In a liquid service the instrument must be below tap. In a gas service the instrument must be above tap. This information should be listed in the Piping ISOs or the P&IDs. Your instrument installation details should match this. If not, engineering must be notified. It is always good practice to follow the instrument installation details when installing a device. Some instruments can be hard mounted. Hard mounting is installing the instrument directly to the pipe or vessel. This should only be done in low vibration areas unless a vibration resistant instrument is used, such as an oil filled pressure indicator, which is most common. In high vibration areas, the instrument should be mounted on a stand and tubing or flex should be used to connect the instrument to the tap. This will give the instrument a longer life. It is always good practice to install a block and bleed with every instrument for future service, no matter how insignificant the instrument may be. A root valve should also be installed at the pipe or vessel to allow a secondary point of isolation for removal and calibration. Note: Try to keep the metals the same. Some metals will not seal well with others, such as carbon steel and stainless steel. Minimize contact between these metals. Always use a Teflon enriched pipe dope or Teflon tape on stainless steel connections.

The instrument should be capable of 120% to 130% of the maximum service pressure. This will protect the instrument from damage from over pressure and prevent faulty readings. Be sure that the pipe, tubing, and fittings are of the proper pressure rating for the install. They should at least be capable of 120% to 130% of the maximum service pressure.

TUBE O.D	STAINLESS STEEL TUBE WALL THICKNESS IN INCHES										
	0.028	0.035	0.049	0.065	0.083	0.095	0.109	0.120	0.134	0.156	0.188
	WORKING PRESSURE IN PSI										
1/8	8500	10900									
3/16	5400	7000	10200								
1/4	4000	5100	7500	10200							
5/16		4000	5800	8000							
3/8		3300	4800	6500	7500						
1/2		2600	3700	5100	6700						
5/8			2900	4000	5200	6000					
3/4			2400	3300	4200	4900	5800				
7/8			2000	2800	3600	4200	4800				
1				2400	3100	3600	4200	4700			
1 1/4					2400	2800	3300	3600	4100	4900	
1 1/2						2300	2700	3000	3400	4000	4900
2							2000	2200	2500	2900	3600

TUBE O.D	COPPER TUBE WALL THICKNESS IN INCHES								
	0.028	0.035	0.049	0.065	0.083	0.095	0.109	0.120	0.134
	WORKING PRESSURE IN PSI								
1/8	2700	3600							
3/16	1800	2300	3400						
1/4	1300	1600	2500	3500					
5/16		1300	1900	2700					
3/8		1000	1600	2200					
1/2		800	1100	1600	2100				
5/8			900	1200	1600	1900			
3/4			700	1000	1300	1500	1800		
7/8			600	800	1100	1300	1500		
1			500	700	900	1100	1300	1500	
1 1/8				600	800	1000	1100	1300	1400

RATING OF STANDARD SCHEDULE 40 STEEL PIPE		
SIZE In.	TYPE	PSI RATING
1/8	CONTINUOUS WELD	700
1/4	CONTINUOUS WELD	700
3/8	CONTINUOUS WELD	700
1/2	CONTINUOUS WELD	700
3/4	CONTINUOUS WELD	700
1	CONTINUOUS WELD	700
1 1/4	CONTINUOUS WELD	800
1 1/2	CONTINUOUS WELD	800
2	CONTINUOUS WELD	800
2 1/2	CONTINUOUS WELD	800
3	CONTINUOUS WELD	800
3 1/2	CONTINUOUS WELD	1200
4	CONTINUOUS WELD	1200
2	ELECTRIC WELD	1000-1300
2 1/2	ELECTRIC WELD	1000-1300
3	ELECTRIC WELD	1000-1300
3 1/2	ELECTRIC WELD	1000-1300
4	ELECTRIC WELD	1000-1300
5	ELECTRIC WELD	1000-1300
6	ELECTRIC WELD	1000-1300
8	ELECTRIC WELD	1000-1300
10	ELECTRIC WELD	1000-1300
12	ELECTRIC WELD	1000-1300
1/8	SEAMLESS	700
1/4	SEAMLESS	700
1/2	SEAMLESS	700
3/4	SEAMLESS	700
1	SEAMLESS	700
1 1/4	SEAMLESS	1000
1 1/2	SEAMLESS	1000
2	SEAMLESS	1000
2 1/2	SEAMLESS	1000
3	SEAMLESS	1000
3 1/2	SEAMLESS	1000-1300
4	SEAMLESS	1000-1300
5	SEAMLESS	1000-1300
6	SEAMLESS	1000-1300
8	SEAMLESS	1000-1300
10	SEAMLESS	1000-1300
12	SEAMLESS	1000-1300

NOTE: All pressure ratings listed are approximate. Consult the manufacturer for a precise pressure rating. Ratings may differ from one manufacturer to the next due to quality standards.

TUBING INSTALLATION:

When installing tubing be sure that the tubing is bottomed out in the fitting. If a close bend is needed, be sure you have a long enough strait section before the start of the bend. See chart below.

LENGTH OF STRAITS BEFORE START of BEND	
TUBE OD	LENGTH
1/16	1/2
1/8	23/32
3/16	3/4
1/4	13/16
5/16	7/8
3/8	15/16
1/2	1 3/16
5/8	1 1/4
3/4	1 1/4
7/8	1 5/16
1	1 1/2
1 1/4	2
1 1/2	2 13/32
2	3 1/4

To tighten the fitting properly, snug the nut with fingers. Put a reference mark on the nut and the fitting body. From this point, with a wrench, tighten nut ¾ of a turn for tubing sizes under ¼ inch OD. For ¼ inch tubing tighten the nut one full turn. For 3/8 and ½ inch tubing tighten the nut 1 ¼ turns. This will compress the ferrules to the tubing, giving a leak proof seal. You can also use a gap gauge. If the gauge will not fit between the nut and body the nut is tight. If the gauge will go between the nut and body, the nut needs to be tightened more.

When placing measurements on the tubing, for placement of bends, be sure to subtract the takeoff from the distance between each bend, in relation to the tubing size. For example, for ½ inch tubing, take off 11/16 inch from the measurement for a 90 degree bend. For 3/8 inch tubing, take off 1/16 inch from the measurement for a 45 degree bend. See chart below.

Degrees BEND ANGLE	TUBE OD					
	1/8	1/4	1/4	5/16	3/8	1/2
	BEND RADIUS					
	9/16	9/16	3/4	15/16	15/16	1 1/2
	TAKE OFF IN INCHES					
15	0	0	0	0	0	0
30	0	0	0	0	0	1/16
45	1/16	1/16	1/16	1/16	1/16	1/16
50	1/16	1/16	1/16	1/16	1/16	1/8
55	1/16	1/16	1/16	1/8	1/8	1/8
60	1/16	1/8	1/16	1/8	1/8	3/16
65	1/8	1/8	1/8	3/16	1/8	1/4
70	1/8	1/8	1/8	3/16	3/16	5/16
75	1/8	3/16	3/16	1/4	1/4	3/8
80	3/16	3/16	3/16	5/16	5/16	7/16
85	1/4	1/4	1/4	3/8	3/8	9/16
90	1/4	5/16	5/16	7/16	7/16	11/16

Screw Pipe:

Threaded pipe is tapered so you have to make sure the pipe screws in far enough to make the seal and that the threads go far enough along the length of pipe. See chart below.

THREADED PIPE			
DISTANCE PIPE SCREWS INTO FITTING FOR A TIGHT FIT			
THREAD TAPER 1/16" PER INCH OF PIPE			
Pipe Size	**Threads Inch**	**Thread Length**	**Screw In Distance**
1/8	27	7/16	5/16
1/4	18	5/8	7/16
3/8	18	5/8	7/16
1/2	14	13/16	9/16
3/4	14	13/16	9/16
1	11 1/2	1	11/16
1 1/4	11 1/2	1	11/16
1 1/2	11 1/2	1 1/32	11/16
2	11 1/2	1 1/16	3/4
2 1/2	8	1 9/16	1 1/16
3	8	1 5/8	1 1/8
4	8	1 3/4	1 3/16
6	8	1 15/16	1 3/8
8	8	2 3/16	1 7/16
10	8	2 3/8	1 5/8
12	8	2 9/16	1 3/4

When threading the pipe, make sure the threads are clean and sharp. Bad threads can create a leak. If the threader will not make clean threads there could be a problem with the die or inadequate lubrication during threading. When pulling measurements for your pipe, subtract the take off of the fittings at each end of the pipe. To find your take off, measure from the center of the cross section of pipe to the edge of the fitting. Subtract the screw in distance of the pipe from this measurement. This will give you your take off. NOTE: The take off for 90's and T's are not always the same. Different manufacturers can also make a difference.

Squares, Cubes, Square Roots, & Cube Roots				
Number	Square	Cube	Square Rt.	Cube Rt.
1	1	1	1.00000	1.00000
2	4	8	1.41421	1.25992
3	9	27	1.73205	1.44225
4	16	64	2.00000	1.58740
5	25	125	2.23607	1.70998
6	36	216	2.44949	1.81712
7	49	343	2.64575	1.91293
8	64	512	2.82843	2.00000
9	81	729	3.00000	2.08008
10	100	1000	3.16228	2.15443

Number	Square	Cube	Square Rt.	Cube Rt.
11	121	1331	3.31662	2.22398
12	144	1728	3.46410	2.28943
13	169	2197	3.60555	2.35133
14	196	2744	3.74166	2.41014
15	225	3375	3.87298	2.46621
16	256	4096	4.00000	2.46621
17	289	4913	4.12311	2.57128
18	324	5832	4.24264	2.62074
19	361	6859	4.35890	2.66840
20	400	8000	4.47214	2.71442
21	441	9261	4.58258	2.75892
22	484	10648	4.69042	2.80204
23	529	12167	4.79583	2.84387
24	576	13824	4.89898	2.88450
25	625	15625	5.00000	2.92402
26	676	17576	5.09902	2.96250
27	729	19683	5.19615	3.00000
28	784	21952	5.29150	3.03659
29	841	24389	5.38516	3.07232
30	900	27000	5.47723	3.10723
31	961	29791	5.56776	3.14138
32	1024	32768	5.65685	3.17480
33	1089	35937	5.74456	3.20753
34	1156	39304	5.83095	3.23961
35	1225	42875	5.91608	3.27107
36	1296	46656	6.00000	3.30193
37	1369	50653	6.08276	3.33222
38	1444	54872	6.16441	3.36198
39	1521	59319	6.24500	3.39121
40	1600	64000	6.32456	3.41995

Number	Square	Cube	Square Rt.	Cube Rt.
41	1681	68921	6.40312	3.44822
42	1764	74088	6.48074	3.47603
43	1849	79507	6.55744	3.50340
44	1936	85184	6.63325	3.53035
45	2025	91125	6.70820	3.55689
46	2116	97336	6.78233	3.58305
47	2209	103823	6.85565	3.60883
48	2304	110592	6.92820	3.63424
49	2401	117649	7.00000	3.65931
50	2500	125000	7.07107	3.68403
51	2601	132651	7.14143	3.70843
52	2704	140608	7.21110	3.73251
53	2809	148877	7.28011	3.75629
54	2916	157464	7.34847	3.77976
55	3025	166375	7.41620	3.80285
56	3136	175616	7.48331	3.82586
57	3249	185193	7.54983	3.84850
58	3364	195112	7.61577	3.87088
59	3481	205379	7.68115	3.89300
60	3600	216000	7.74597	3.91487
61	3721	226981	7.81025	3.93650
62	3844	238328	7.87401	3.95789
63	3969	250047	7.93725	3.97906
64	4096	262144	8.00000	4.00000
65	4225	274625	8.06226	4.02073
66	4356	287496	8.12404	4.04124
67	4489	300763	8.18535	4.06155
68	4624	314432	8.24621	4.08166
69	4761	328509	8.30662	4.10157
70	4900	343000	8.36660	4.12129

Number	Square	Cube	Square Rt.	Cube Rt.
71	5041	357911	8.42615	4.14082
72	5184	373248	8.48528	4.16017
73	5329	389017	8.54400	4.17934
74	5476	405224	8.60233	4.19834
75	5625	421875	8.66025	4.21716
76	5776	438976	8.71780	4.23582
77	5929	456533	8.77496	4.25432
78	6084	474552	8.83176	4.27266
79	6241	493039	8.88819	4.29084
80	6400	512000	8.94427	4.30887
81	6561	531441	9.00000	4.32675
82	6724	551368	9.05539	4.34448
83	6889	571787	9.11043	4.36207
84	7056	592704	9.16515	4.37952
85	7225	614125	9.21954	4.39683
86	7396	636056	9.27362	4.41400
87	7569	658503	9.32738	4.43105
88	7744	681472	9.38083	4.44796
89	7921	704969	9.43398	4.48475
90	8100	729000	9.48683	4.48140
91	8281	753571	9.53939	4.49794
92	8464	778688	9.59166	4.51436
93	8649	804357	9.64365	4.53065
94	8836	830584	9.69536	4.54684
95	9025	857375	9.74679	4.56290
96	9216	884736	9.79796	4.57886
97	9409	912673	9.84886	4.59470
98	9604	941192	9.89949	4.61044
99	9801	970299	9.94987	4.62607
100	10000	1000000	10.00000	4.64159

DEGREES & TRIG FUNCTIONS

n	n Radions	Sine	Cosine	Tangent
0	0.00000	0.00000	1.00000	0.00000
1	0.01745	0.01745	0.99985	0.01746
2	0.03491	0.03490	0.99939	0.03492
3	0.05236	0.05234	0.99863	0.05241
4	0.06981	0.06976	0.99756	0.06993
5	0.08727	0.08716	0.99619	0.08749
6	0.10472	0.10453	0.99452	0.10510
7	0.12217	0.12187	0.99255	0.12178
8	0.13963	0.13917	0.99027	0.14054
9	0.15708	0.15643	0.98769	0.15838
10	0.17453	0.17365	0.98481	0.17633
11	0.19199	0.19081	0.98163	0.19438
12	0.20944	0.20791	0.97815	0.21256
13	0.22689	0.22495	0.97437	0.23087
14	0.24435	0.24192	0.97030	0.24933
15	0.26180	0.25882	0.96593	0.26795
16	0.27925	0.27564	0.96126	0.28675
17	0.29671	0.29237	0.95630	0.30573
18	0.31416	0.30902	0.95106	0.32492
19	0.33161	0.32557	0.94552	0.34433
20	0.34907	0.34202	0.93969	0.36397
21	0.36652	0.35837	0.93358	0.38386
22	0.38397	0.37461	0.92718	0.40403
23	0.40143	0.39073	0.92050	0.42447
24	0.41888	0.40674	0.91355	0.44523
25	0.43633	0.42262	0.90631	0.46631
26	0.45379	0.43837	0.89879	0.48773
27	0.47124	0.45399	0.89101	0.50953
28	0.48869	0.46947	0.88295	0.53171
29	0.50615	0.48481	0.87462	0.55431
30	0.52360	0.50000	0.86603	0.57735

n	n Radions	Sine	Cosine	Tangent
31	0.54105	0.51504	0.85717	0.60086
32	0.55851	0.52992	0.84805	0.62487
33	0.57596	0.54464	0.83867	0.64941
34	0.59341	0.55919	0.82904	0.67451
35	0.61087	0.57358	0.81915	0.70021
36	0.62832	0.58779	0.80902	0.72654
37	0.64577	0.60182	0.79864	0.75355
38	0.66323	0.61566	0.78801	0.78129
39	0.68068	0.62932	0.77715	0.80978
40	0.69813	0.64279	0.76604	0.83910
41	0.71558	0.65606	0.75471	0.86929
42	0.73304	0.66913	0.74314	0.90040
43	0.75049	0.68200	0.73135	0.93252
44	0.76794	0.69466	0.71934	0.96569
45	0.78540	0.70711	0.70711	1.00000
46	0.80285	0.71934	0.69466	1.03553
47	0.82030	0.73135	0.68200	1.07237
48	0.83776	0.74314	0.66913	1.11061
49	0.85521	0.75471	0.65606	1.15037
50	0.87266	0.76604	0.64279	1.19175
51	0.89012	0.77715	0.62932	1.23490
52	0.90757	0.78801	0.61566	1.27994
53	0.92502	0.79864	0.60182	1.32704
54	0.94248	0.80902	0.58779	1.37638
55	0.95993	0.81915	0.57358	1.42815
56	0.97738	0.82904	0.55919	1.48256
57	0.99484	0.83867	0.54464	1.53986
58	1.01229	0.84805	0.52992	1.60033
59	1.02974	0.85717	0.51504	1.64428
60	1.04720	0.86603	0.50000	1.73205

n	n Radions	Sine	Cosine	Tangent
61	1.06465	0.87462	0.48481	1.80405
62	1.08210	0.88295	0.46947	1.88073
63	1.09956	0.89101	0.45399	1.96261
64	1.11701	0.89879	0.43837	2.05030
65	1.13446	0.90631	0.42262	2.14451
66	1.15192	0.91355	0.40674	2.24604
67	1.16937	0.92050	0.39073	2.35585
68	1.18682	0.92718	0.37461	2.47509
69	1.20428	0.93358	0.35837	2.60509
70	1.22173	0.93969	0.34202	2.74748
71	1.23918	0.94552	0.32557	2.90421
72	1.25664	0.95106	0.30902	3.07768
73	1.27409	0.95630	0.29237	3.27085
74	1.29154	0.96126	0.27564	3.48741
75	1.30900	0.96593	0.25882	3.73205
76	1.32645	0.97030	0.24192	4.01078
77	1.34390	0.97437	0.22495	4.33148
78	1.36136	0.97815	0.20791	4.70463
79	1.37881	0.98163	0.19081	5.14455
80	1.39626	0.98481	0.17365	5.67128
81	1.41372	0.98769	0.15643	6.31375
82	1.43117	0.99027	0.13917	7.11537
83	1.44862	0.99255	0.12187	8.14435
84	1.46608	0.99452	0.10453	9.51436
85	1.48353	0.99619	0.08716	11.43005
86	1.50098	0.99756	0.06976	14.30067
87	1.51844	0.99863	0.05234	19.08114
88	1.53589	0.99939	0.03490	28.63625
89	1.55334	0.99985	0.01745	57.28996
90	1.57080	1.00000	0.00000	infinity

n	n Radions	Sine	Cosine	Tangent
91	1.58825	0.99985	-0.01745	-57.28996
92	1.60570	0.99939	-0.03490	-28.63625
93	1.62316	0.99863	-0.05234	-19.08114
94	1.64061	0.99756	-0.06976	-14.30067
95	1.65806	0.99619	-0.08716	-11.43005
96	1.67552	0.99452	-0.10453	-9.51436
97	1.69297	0.99255	-0.12187	-8.14435
98	1.71042	0.99027	-0.13917	-7.11537
99	1.72788	0.98769	-0.15643	-6.31375
100	1.74533	0.98481	-0.17365	-5.67128
101	1.76278	0.98163	-0.19081	-5.14455
102	1.78024	0.97815	-0.20791	-4.70463
103	1.79769	0.97437	-0.22495	-4.33148
104	1.81514	0.97030	-0.24192	-4.01078
105	1.83260	0.96593	-0.25882	-3.73205
106	1.85005	0.96126	-0.27564	-3.48741
107	1.86750	0.95630	-0.29237	-3.27085
108	1.88496	0.95106	-0.30902	-3.07768
109	1.90241	0.94552	-0.32557	-2.90421
110	1.91986	0.93969	-0.34202	-2.74748
111	1.93732	0.93358	-0.35837	-2.60509
112	1.95477	0.92718	-0.37461	-2.47509
113	1.97222	0.92050	-0.39073	-2.35585
114	1.98968	0.91355	-0.40674	-2.24604
115	2.00713	0.90631	-0.42262	-2.14451
116	2.02458	0.89879	-0.43837	-2.05030
117	2.04204	0.89101	-0.45399	-1.96261
118	2.05949	0.88295	-0.46947	-1.88073
119	2.07694	0.87462	-0.48481	-1.80405
120	2.09440	0.86603	-0.50000	-1.73205

n	n Radions	Sine	Cosine	Tangent
121	2.11185	0.85717	-0.51504	-1.66428
122	2.12930	0.84805	-0.52992	-1.60033
123	2.14675	0.83867	-0.54464	-1.53986
124	2.16421	0.82904	-0.55919	-1.48256
125	2.18166	0.81915	-0.57358	-1.42815
126	2.19911	0.80902	-0.58779	-1.37638
127	2.21657	0.79864	-0.60182	-1.32704
128	2.23402	0.78801	-0.61566	-1.27994
129	2.25147	0.77715	-0.62932	-1.23490
130	2.26893	0.76604	-0.64279	-1.19175
131	2.28638	0.75471	-0.65606	-1.15037
132	2.30383	0.74314	-0.66913	-1.11061
133	2.32129	0.73135	-0.68200	-1.07237
134	2.33874	0.71934	-0.69466	-1.03553
135	2.35619	0.70711	-0.70711	-1.00000
136	2.37365	0.69466	-0.71934	-0.96569
137	2.39110	0.68200	-0.73135	-0.93252
138	2.40855	0.66913	-0.74314	-0.90040
139	2.42601	0.65606	-0.75471	-0.86929
140	2.44346	0.64279	-0.76604	-0.83910
141	2.46091	0.62932	-0.77715	-0.80978
142	2.47837	0.61566	-0.78801	-0.78129
143	2.49582	0.60182	-0.79864	-0.75355
144	2.51327	0.58779	-0.80902	-0.72654
145	2.53073	0.57358	-0.81915	-0.70021
146	2.54818	0.55915	-0.82904	-0.67451
147	2.56563	0.54464	-0.83867	-0.64941
148	2.58309	0.52992	-0.84805	-0.62487
149	2.60054	0.51504	-0.85717	-0.60086
150	2.61799	0.50000	-0.86603	-0.57735

n	n Radions	Sine	Cosine	Tangent
151	2.63545	0.48481	-0.87462	-0.55431
152	2.65290	0.46947	-0.88295	-0.53171
153	2.67035	0.45399	-0.89101	-0.50953
154	2.68781	0.43837	-0.89879	-0.48773
155	2.70526	0.42262	-0.90631	-0.46631
156	2.72271	0.40674	-0.91355	-0.44523
157	2.74017	0.39073	-0.92050	-0.42447
158	2.75762	0.37461	-0.92718	-0.40403
159	2.77507	0.35837	-0.93358	-0.38386
160	2.79253	0.34202	-0.93969	-0.36397
161	2.80998	0.32557	-0.94552	-0.34433
162	2.82743	0.30902	-0.95106	-0.32492
163	2.84489	0.29237	-0.95630	-0.30573
164	2.86234	0.27564	-0.96126	-0.28675
165	2.87979	0.25882	-0.96593	-0.26795
166	2.89725	0.24192	-0.97030	-0.24933
167	2.91470	0.22495	-0.97437	-0.23087
168	2.93215	0.20791	-0.97815	-0.21256
169	2.94961	0.19081	-0.98163	-0.19438
170	2.96706	0.17365	-0.98481	-0.17633
171	2.98451	0.15643	-0.98769	-0.15838
172	3.00197	0.13917	-0.99027	-0.14054
173	3.01942	0.12187	-0.99255	-0.12278
174	3.03687	0.10453	-0.99452	-0.10510
175	3.05433	0.08716	-0.99619	-0.08749
176	3.07178	0.06976	-0.99756	-0.06993
177	3.08923	0.05234	-0.99863	-0.05241
178	3.10669	0.03490	-0.99939	-0.03492
179	3.12414	0.01745	-0.99985	-0.01746
180	3.14159	0.00000	-1.00000	0.00000

n	n Radions	Sine	Cosine	Tangent
181	3.15905	-0.01745	-0.99985	0.01746
182	3.17650	-0.03490	-0.99939	0.03492
183	3.19395	-0.05234	-0.99863	0.05241
184	3.21141	-0.06976	-0.99756	0.06993
185	3.22886	-0.08716	-0.99619	0.08749
186	3.24631	-0.10453	-0.99452	0.10510
187	3.26377	-0.12187	-0.99255	0.12278
188	3.28122	-0.13917	-0.99027	0.14054
189	3.29867	-0.15643	-0.98769	0.15838
190	3.31613	-0.17365	-0.98481	0.17633
191	3.33358	-0.19081	-0.98163	0.19438
192	3.35103	-0.20791	-0.97815	0.21256
193	3.36849	-0.22495	-0.97437	0.23087
194	3.38594	-0.24192	-0.97030	0.24933
195	3.40339	-0.25882	-0.96593	0.26795
196	3.42085	-0.27564	-0.96126	0.28675
197	3.43830	-0.29237	-0.95630	0.30573
198	3.45575	-0.30902	-0.95106	0.32492
199	3.47321	-0.32557	-0.94552	0.34433
200	3.49066	-0.34202	-0.93969	0.36397
201	3.50811	-0.35837	-0.93358	0.38386
202	3.52557	-0.37461	-0.92718	0.40403
203	3.54302	-0.39073	-0.92050	0.42447
204	3.56047	-0.40674	-0.91355	0.44523
205	3.57792	-0.42262	-0.90631	0.46631
206	3.59538	-0.43837	-0.89879	0.48773
207	3.61283	-0.45399	-0.89101	0.50953
208	3.63028	-0.46947	-0.88295	0.53171
209	3.64774	-0.48481	-0.87462	0.55431
210	3.66519	-0.50000	-0.86603	0.57735

n	n Radions	Sine	Cosine	Tangent
211	3.68264	-0.51504	-0.85717	0.60086
212	3.70010	-0.52992	-0.84805	0.62487
213	3.71755	-0.54464	-0.83867	0.64941
214	3.73500	-0.55919	-0.82904	0.67451
215	3.75246	-0.57358	-0.81915	0.70021
216	3.76991	-0.58779	-0.80902	0.72654
217	3.78736	-0.60182	-0.79864	0.75355
218	3.80482	-0.61566	-0.78801	0.78129
219	3.82227	-0.62932	-0.77715	0.80978
220	3.83972	-0.64279	-0.76604	0.83910
221	3.85718	-0.65606	-0.75471	0.86929
222	3.87463	-0.66913	-0.74314	0.90040
223	3.89208	-0.68200	-0.73135	0.93252
224	3.90954	-0.69466	-0.71934	0.96569
225	3.92699	-0.70711	-0.70711	1.00000
226	3.94444	-0.71934	-0.69466	1.03553
227	3.96190	-0.73135	-0.68200	1.07237
228	3.97935	-0.74314	-0.66913	1.11061
229	3.99680	-0.75471	-0.65606	1.15037
230	4.01426	-0.76604	-0.64279	1.19175
231	4.03171	-0.77715	-0.62932	1.23490
232	4.04916	-0.78801	-0.61566	1.27994
233	4.06662	-0.79864	-0.60182	1.32704
234	4.08407	-0.80902	-0.58779	1.37638
235	4.10152	-0.81915	-0.57358	1.42815
236	4.11898	-0.82904	-0.55919	1.48256
237	4.13643	-0.83867	-0.54464	1.53986
238	4.15388	-0.84805	-0.52992	1.60033
239	4.17134	-0.85717	-0.51504	1.66428
240	4.18879	-0.86603	-0.50000	1.73205

n	n Radions	Sine	Cosine	Tangent
241	4.20624	-0.87462	-0.48481	1.80405
242	4.22370	-0.88295	-0.46947	1.88073
243	4.24115	-0.89101	-0.45399	1.96261
244	4.25860	-0.89879	-0.43837	2.05030
245	4.27606	-0.90631	-0.42262	2.14451
246	4.29351	-0.91355	-0.40674	2.24604
247	4.31096	-0.92050	-0.39073	2.35585
248	4.32842	-0.92718	-0.37461	2.47509
249	4.34587	-0.93358	-0.35837	2.60509
250	4.36332	-0.93969	-0.34202	2.74748
251	4.38078	-0.94552	-0.32557	2.90421
252	4.39823	-0.95106	-0.30902	3.07768
253	4.41568	-0.95630	-0.29237	3.27085
254	4.43140	-0.96126	-0.27564	3.48741
255	4.45059	-0.96593	-0.25882	3.73205
256	4.46804	-0.97030	-0.24192	4.01078
257	4.48550	-0.97437	-0.22495	4.33148
258	4.50295	-0.97815	-0.20791	4.70463
259	4.52040	-0.98163	-0.19081	5.14455
260	4.53786	-0.98481	-0.17365	5.67128
261	4.55531	-0.98769	-0.15643	6.31375
262	4.57276	-0.99027	-0.13917	7.11537
263	4.59022	-0.99255	-0.12187	8.14435
264	4.60767	-0.99452	-0.10453	9.51436
265	4.62512	-0.99619	-0.07816	11.43005
266	4.64258	-0.99756	-0.06976	14.30067
267	4.66003	-0.99863	-0.05234	19.08114
268	4.67748	-0.99939	-0.03490	28.63625
269	4.69494	-0.99985	-0.01745	57.28996
270	4.71239	-1.00000	0.00000	infinity

n	n Radions	Sine	Cosine	Tangent
271	4.72984	-0.99985	0.01745	-57.28996
272	4.74730	-0.99939	0.03490	-28.63625
273	4.76475	-0.99863	0.05234	-19.08114
274	4.78220	-0.99756	0.06976	-14.30067
275	4.79966	-0.99619	0.08716	-11.43005
276	4.81711	-0.99452	0.10453	-9.51436
277	4.83456	-0.99255	0.12187	-8.14435
278	4.85202	-0.99027	0.13917	-7.11537
279	4.86947	-0.98769	0.15643	-6.31375
280	4.88692	-0.98481	0.17365	-5.67128
281	4.90438	-0.98163	0.19081	-5.14455
282	4.92183	-0.97815	0.20791	-4.70463
283	4.93928	-0.97437	0.22495	-4.33148
284	4.95674	-0.97030	0.24192	-4.01078
285	4.97419	-0.96593	0.25882	-3.73205
286	4.99164	-0.96126	0.27564	-3.48741
287	5.00909	-0.95630	0.29237	-3.27085
288	5.02655	-0.95106	0.30902	-3.07768
289	5.04400	-0.94552	0.32557	-2.90421
290	5.06145	-0.93969	0.34202	-2.74748
291	5.07891	-0.93358	0.35837	-2.60509
292	5.09636	-0.92718	0.37461	-2.47509
293	5.11381	-0.92050	0.39073	-2.35585
294	5.13127	-0.91355	0.40674	-2.24604
295	5.14872	-0.90631	0.42262	-2.14451
296	5.16617	-0.89879	0.43837	-2.05030
297	5.18363	-0.89101	0.45399	-1.96261
298	5.20108	-0.88295	0.46947	-1.88073
299	5.21853	-0.87462	0.48481	-1.80405
300	5.23599	-0.86603	0.50000	-1.73205

n	n Radions	Sine	Cosine	Tangent
301	5.25344	-0.85717	0.51504	-1.66428
302	5.27089	-0.84805	0.52992	-1.60033
303	5.28835	-0.83867	0.54464	-1.53986
304	5.30580	-0.82904	0.55919	-1.48256
305	5.32325	-0.81915	0.57358	-1.42815
306	5.34071	-0.80902	0.58779	-1.37638
307	5.35816	-0.79864	0.60182	-1.32704
308	5.37561	-0.78801	0.61566	-1.27994
309	5.39307	-0.77715	0.62932	-1.23490
310	5.41052	-0.76604	0.64279	-1.19175
311	5.42797	-0.75471	0.65606	-1.15037
312	5.44543	-0.74314	0.66913	-1.11061
313	5.46288	-0.73135	0.68200	-1.07237
314	5.48033	-0.71934	0.69466	-1.03553
315	5.49779	-0.70711	0.70711	-1.00000
316	5.51524	-0.69466	0.71934	-0.96569
317	5.53269	-0.68200	0.73135	-0.93252
318	5.55015	-0.66913	0.74314	-0.90040
319	5.56760	-0.65606	0.75471	-0.86929
320	5.58505	-0.64279	0.76604	-0.83910
321	5.60251	-0.62932	0.77715	-0.80978
322	5.61996	-0.61566	0.78801	-0.78129
323	5.63741	-0.60182	0.79864	-0.75355
324	5.65487	-0.58779	0.80902	-0.72654
325	5.67232	-0.57358	0.81915	-0.70021
326	5.68977	-0.55919	0.82904	-0.67451
327	5.70723	-0.54464	0.83867	-0.64941
328	5.72468	-0.52992	0.84805	-0.62487
329	5.74213	-0.51504	0.85717	-0.60086
330	5.75959	-0.50000	0.86603	-0.57735

n	n Radions	Sine	Cosine	Tangent
331	5.77704	-0.48481	0.87462	-0.55431
332	5.79449	-0.46947	0.88295	-0.53171
333	5.81195	-0.45399	0.89101	-0.50953
334	5.82940	-0.43837	0.89879	-0.48773
335	5.84685	-0.42262	0.90631	-0.46631
336	5.86431	-0.40674	0.91355	-0.44523
337	5.88176	-0.39073	0.92050	-0.42447
338	5.89921	-0.37461	0.92718	-0.40403
339	5.91667	-0.35837	0.93358	-0.38386
340	5.93412	-0.34202	0.93969	-0.36397
341	5.95157	-0.32557	0.94552	-0.34433
342	5.96903	-0.30902	0.95106	-0.32492
343	5.98648	-0.29237	0.95630	-0.30573
344	6.00393	-0.27564	0.96126	-0.28675
345	6.02139	-0.25882	0.96593	-0.26795
346	6.03884	-0.24192	0.97030	-0.24933
347	6.05629	-0.22495	0.97437	-0.23087
348	6.07375	-0.20791	0.97815	-0.21256
349	6.09120	-0.19081	0.98163	-0.19438
350	6.10865	-0.17365	0.98481	-0.17633
351	6.12611	-0.15643	0.98769	-0.15838
352	6.14356	-0.13917	0.99027	-0.14054
353	6.16101	-0.12187	0.99255	-0.12278
354	6.17847	-0.10453	0.99452	-0.10510
355	6.19592	-0.08716	0.99619	-0.08749
356	6.21337	-0.06976	0.99756	-0.06993
357	6.23083	-0.05234	0.99863	-0.05241
358	6.24828	-0.03490	0.99939	-0.03492
359	6.26573	-0.01745	0.99985	-0.01746
360	6.28319	0.00000	1.00000	0.00000

Trig Formulas

FINDING FUNCTIONS OF ANGLES	
SINE =	SIDE OPPOSITE / HYPOTENUSE
COSINE =	SIDE ADJACENT / HYPOTENUSE
TANGENT =	SIDE OPPOSITE / SIDE ADJACENT
COTANGENT =	SIDE ADJACENT / SIDE OPPOSITE
SECANT =	HYPOTENUSE / SIDE ADJACENT
COSECANT =	HYPOTENUSE / SIDE OPPOSITE

Right Triangle

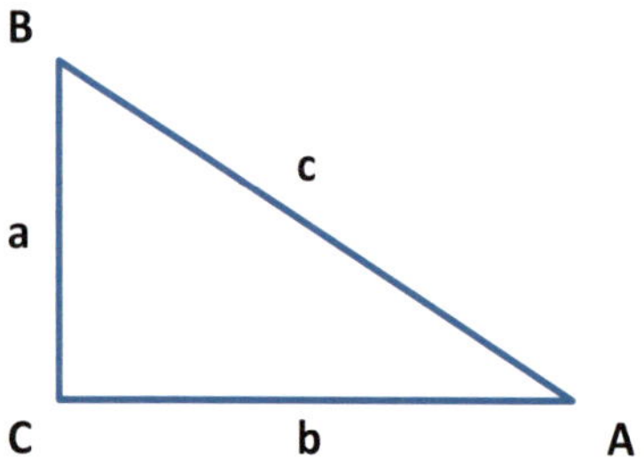

A,B,C = Angles *a,b,c = Sides*

Note: Side a is adjacent to angle B. Side b is adjacent to angle A. Side c is the hypotenuse. Side a is opposite angle A. Side b is opposite angle B. Side c is opposite angle C.

Angles:

sine A = a/c = cosine B

cosine A = b/c = sine B

tangent A = a/b = cotangent B

cotangent A = b/a = tangent B

secant A = c/b = cosecant B

cosecant A = c/a = secant B

Area = (ab)/2

Sides:

Length of Side Opposite:

Hypotenuse **X** Sine,
Hypotenuse/Cosecant,

Side Adjacent **X** Tangent,
Side Adjacent **/** Cotangent

Length of Side Adjacent:

Hypotenuse **X** Cosine,
Hypotenuse **/** Secant,
Side Opposite **X** Cotangent,
Side Opposite **/** Tangent

Length of Hypotenuse:

Side Opposite **X** Cosecant,
Side Opposite **/** Secant,
Side Adjacent **X** Secant,
Side Adjacent **/** Cosine

a = b **X** cot A = c **X** sin A
b = a **X** cot A = c **X** cos A
c = a **/** sin A = b **/** cos A

Oblique Triangle

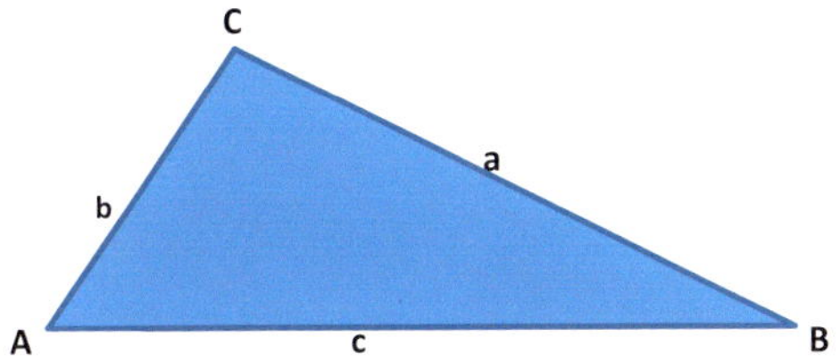

a = b sinA / sinB b = a sinB / sinA c = a sinC / sinA

A = 180° – (B+C) B = 180° – (A+C) C = 180° – (A+B)

Sine A = a sinB / b Sine B = b sinA / a Cosine A = $b^2 + c^2 – a^2$ / 2bc Cosine B = $a^2 + c^2 – b^2$ / 2ac

Tangent A = a sinC / b – (a cosC) Tangent B = b sinC / a – (b cosC)

Area = bc sinA / 2 = a^2 sinB sinC / 2 sinA

Equilateral Triangle

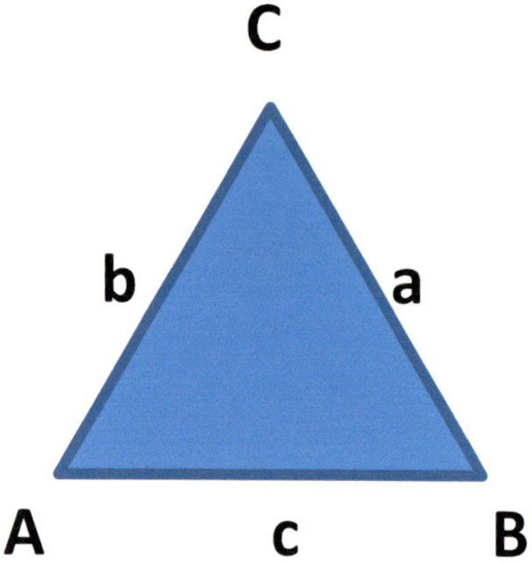

All sides are equal and all angles are equal. All angles are 60°.

A = B = C and a = b = c

Height = any side **X** 0.866

Area = any side^2 **X** 0.433

Polygons

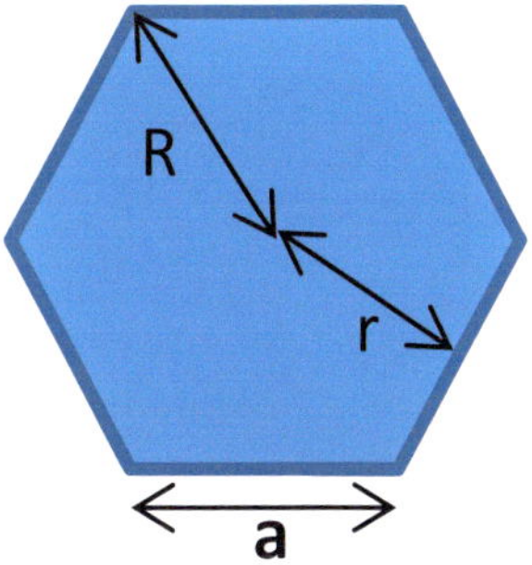

n = number of sides (all sides equal)

a = length of sides

r = radius to center of side

R = radius to center of angle

θ = degrees

A = area

P = perimeter

Perimeter = na

Area = nar / 2 = $nr^2 \tan\theta$ = $(n R^2 / 2)\sin 2\theta$

Annulus

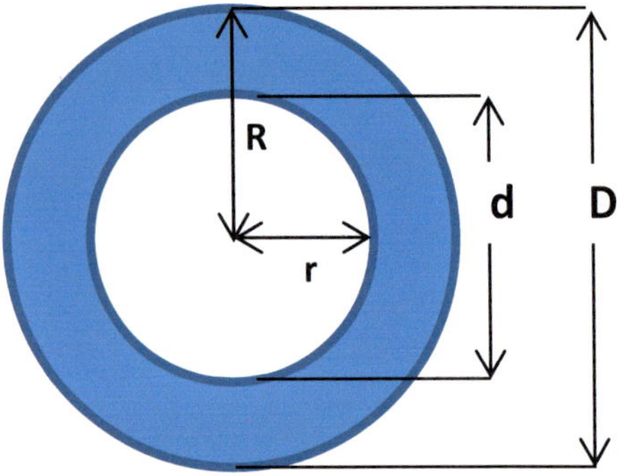

d = inside diameter

D = outside diameter

r = inside radius

R = outside radius

c = circumference

Area = $(D^2 - d^2)0.7854 = \pi\,(r + R)(R - r)$

Circumference d = $2\pi r = \pi d$

Circumference D = $2\pi R = \pi D$

To find the circumference for the center of an annulus: $\pi((d + D)/2)$. To find the Diameter of the annulus from centers: $(d+D)/2$. These are helpful when bending tubing, conduit, or fitting pipe.

Resistors

When resistors are in series, the total resistance is the sum of all the resistors. **Rtotal** = $R_1 + R_2 + R_3$....etc.

When resistors are in parallel the total resistance is equal to the sum of the reciprocals of the resistances. **(1 / Rtotal)** = $1 / ((1/R_1) + (1/R_2) + (1/R_3) +$...etc.)

MHO

The mho is the unit of conductance. It is the reciprocal of resistance. The relation between conductance G and resistance R is given by G = I/R. Conductivity is the ease at which a current may be forced through a circuit. Resistance is the difficulty with which a current if forced though a circuit.

RESISTOR COLOR CODES					
BAND 1	BAND2	BAND3	BAND4	BAND5	BAND6
			X0.1	5%	
			X0.01	10%	
0	0	0	X1		
1	1	1	X10	1%	100ppm
2	2	2	X100	2%	50ppm
3	3	3	X1,000		15ppm
4	4	4	X10,000		25ppm
5	5	5	X100,000	0.50%	
6	6	6	X1,000,000	0.25%	
7	7	7	X10,000,000	0.10%	
8	8	8	X100,000,000		
9	9	9	X1,000,000,000		
FIRST DIGIT	SECOND DIGIT	THIRD DIGIT	MULTIPLIER	TOLERANCE	Temp. Coefficient
Craig Jay Riley		Five and Six band only	Band3 on Four band resistor		Six band resistor only

Capacitors

When capacitors are in series, the total capacitance is equal to the sum of the reciprocals of the capacitors. **(1 / Ctotal)** = 1 / ((1/ C_1) + (1/C_2) + (1/C_3) + ...etc.)

When capacitors are in parallel, the total capacitance is the sum of all the capacitors. **Ctotal** = $C_1 + C_2 + C_3 + \ldots$.etc.

Inductors

E = voltage, **L** = inductor, **I** = current

In an inductive circuit, voltage leads current by 90°. "ELI"

When inductors are in series, the total inductance is the sum of all the inductors. **Ltotal** = $L_1 + L_2 + L_3 + \ldots$etc.

When inductors are in parallel, the total inductance is equal to the sum of the reciprocals of the inductors. (1 / Ltotal) = 1 / ((1/L_1) + (1/L_2) +(1/L_3) + ...etc.)

Ohm's Law

E = voltage = IR = P/I = $\sqrt{PR}$

I = current or "amperage" = E/R = P/E = $\sqrt{\frac{P}{R}}$

R = resistance "ohms" = $E/I = E^2/P = P/I^2$

P = Power "watts" = $EI = I^2R = (IR)I = E^2/R$

Pt = Power total = $P_1 + P_2 + P_3 + ...$etc.

Analog Calculations

These formulas will allow you to calculate values for analog devices, even devices having a lower range not equal to zero.

Ispan = **I**max – **I**min = 16ma

ma per unit = span/(lower range + upper range)

ma = ((value – lower range) ma per unit) + 4ma

Units per ma = (lower range + upper range)/**I**span

Unit = ((**I**value – 4ma) units per ma) +lower range

Formulas

Pi = 3.1415926

Head Pressure psi = (height "ft")(specific gravity)0.43307

Head Pressure KPa = (height "meters")(specific gravity)9.1012

Flow Velocity ft/second = (gallons/minute)/(π(pipe diameter/2)^2)3.209

Flow Velocity m/second = (liters/minute)/(π(pipe diameter/2)^2)38.704

Flow Velocity = Flow rate**/**Area

Flow Rate = Flow velocity **X** Flow Area

Power "kW" = (Flow rate **X** Pressure)/60

Hydraulic Horsepower "HHP" = (Rate **X** Pressure)/256.8 = Power "kW"/0.746

°Celsius "°C" to °Fahrenheit "°F" = (°C **X** 1.8)+32°

°Fahrenheit "°F" to °Celsius "°C" = (°F-32°)/1.8

Transmitters

Pressure transmitters are only used for measuring pressure.

Differential pressure transmitters can be used for pressure, differential pressure, flow, or level.

When used as a pressure transmitter, simply tie the high side in and vent the low side. For example, if used as a damper control, you may have to block the wind from the vented side because the pressure of the wind will throw off your calibration. Also if you are using barometric pressure ("Hg), call the local airport or an i-phone and get the current barometric pressure. Subtract that from your reading with both sides vented to get your zero.

When used as a differential pressure, tie the high side to the upstream and the low side to the downstream. For example, on a filter the high side will be connected on the inlet side of the filter and the low side to the outlet of the filter.

When used as a flow transmitter, use the same principle as for differential pressure. High side is upstream and low side is downstream. Be sure that the orifice plate is installed in the correct direction. Incorrect installation can cause an inaccurate reading on your transmitter. The stamp should be upstream.

When used as a level transmitter, in most cases the high side will be connected to your lower tap and your low side connected to the higher elevation tap. ""WARNING" **Do not zero the transmitter.**" Your lower range value is almost never zero. Most of the time, it will be a negative number. With both sides vented, use the output reading as your lower range value. If you are using wet legs, make sure both legs are filled with product and vented at tap connection to get your lower range value. If there is a difference in height between your lower tap and your lower range height, multiply the difference by the specific gravity and add this to your reading to get your lower range value. To find your upper range value, multiply your range by the specific gravity and add this to your lower range value to get your upper range value.

When used as a reverse reading level, you will tie your high side into the higher elevation tap and your low side into your lower tap. Your lower range value will be greater than your upper range value. This is done because most transmitters have a larger range in the negative than in the positive. This will setup the same as the other except your upper and lower range values will be reversed.

Redundant Transmitters

Differential pressure: When setting up redundant transmitters, make sure that your zero% is at the same point on both transmitters. The value may be different but if you enter the two different values as your lower ranges, you will get a more accurate calibration and they will read the same. For example, when using wet legs, the wet legs will more than likely be different lengths causing a difference in readout. Make sure all legs are filled and vented at taps. This will give you a true zero% on both transmitters. From there you can add your engineered range and get your upper range value.

If used as a level, be sure to multiply your engineered range by the specific gravity of the process liquid to get your range. Then add this to your lower range value to get your upper range value.

Capillaries: Capillaries are a type of differential pressure. Basically, the diaphragm is remote mounted instead of inside the transmitter. Make sure capillaries are mounted at the taps before attempting calibration. The height of the capillaries will make a difference, so bench calibration is out of the question. With redundant capillaries, set up the lower range the same as a differential pressure. If the lower taps are at different heights, add the height difference multiplied by the specific gravity to get your matching lower range value. If your upper taps are at different heights, after your zero% has been achieved, pump up both transmitters to 100% of the range and input each individual reading into the corresponding transmitter. Most likely the values will not be the same. Now you have matching ranges.

Displacers: Measure the distance between the center of your lower tap and the center of your upper tap. This will give you your upper range limit. This must be greater than or equal to the engineered range. The center of your lower tap is your lower range limit. Do not use the bottom of the tap as zero%. Your reading should go a little below zero% when the vessel is dry.

When setting up redundant displacers, your taps deed to be the same height or the range will be limited to the distance between the center of your highest low tap and the center of your lowest high tap. Not following this rule will cause the transmitters to read differently through the scale. It is best to float both at the same time with a poly tubing rig so that you can see your true level. Using the center of the highest low tap, set that reading value as your lower range value on each of your transmitters individually. Float both to the center of the lowest upper tap. Set this reading value as your upper range limit on each transmitter individually. Now life is good.

I/P

To set up an i/p, using a meter, drive 4ma to the i/p. Adjust the output pressure to 3 psi with the zero pot. Now drive 20ma to the i/p. Adjust the output pressure to 15 psi with the span pot. Using the meter, drive up and down scale to make sure the output stays at 3 to 15 psi. If it doesn't, repeat the steps.

P/I

To setup a p/i, use a regulator to control the air flow. Adjust the pressure to 3 psi. Using a reference meter, read the output ma. Adjust he output to 4ma with the zero pot. Adjust the input pressure to 15 psi with the regulator. Adjust the output to 20ma with the span pot. Using the regulator go up and down scale to check repeatability. Repeat steps if necessary.

Valves: AOV's

To adjust valve stroke, use a regulator to control the air flow to the cylinder and move valve to the closed position. Remove locking nut, to free the stem from shaft. Open the valve to full open for clearance. Place a flat block of wood between the stem and shaft and close valve. This will insure the valve is seated properly. Once valve is seated, open valve, moving stem up, and remove block of wood.

With valve fully opened, place a reference mark on both the stem and actuator body. Measure down from the reference mark on body, the distance the stroke should be and place a second mark on body. Using the regulator, move the actuator stem to match the reference mark on the stem with the second mark on body. Recouple the stem to the shaft with the locking nut. Remove all pressure with the regulator. If this is a fail close valve the valve should remain closed. Using the regulator, slowly adjust the pressure until you feel the shaft start to move. This should happen at 3 psi. If not, adjust the spring nut until it does. Once adjusted, slowly adjust the regulator to 15 psi. The valve should now be at full open. If not the spring nut may have to be readjusted. The valve should be fully seated at 3 psi, full open at 15 psi, and half way at 9psi. Now your valve is set.

COPPER WIRE TABLE

Resistance of pure copper in Ohms at 68 deg. F

Gauge #	Diameter in inches	Area in circular mils	Ohms per 1000 ft.	Feet per Ohm
OOOO	0.46000		0.0490	20408.1
OOO	0.40960		0.0618	16181.2
OO	0.36480		0.0779	12837.0
O	0.32490		0.0983	10172.9
1	0.28930		0.1239	8071.0
2	0.25760		0.1563	6398.0
3	0.22940		0.1970	5076.1
4	0.20430		0.2485	4024.1
5	0.18190		0.3133	3191.8
6	0.16200		0.3951	2531.0
7	0.14430		0.4982	2007.2
8	0.12850		0.6282	1591.8
9	0.11440		0.7921	1262.5
10	0.10190	10380	0.9989	1001.1
11	0.09070	8234	1.2600	793.6
12	0.08081	6530	1.5880	629.7
13	0.07196	5178	2.0030	499.3
14	0.06408	4107	2.5250	396.0
15	0.05707	3257	3.1840	314.1
16	0.05082	2583	4.0160	249.0
17	0.04526	2048	5.0640	197.5
18	0.04030	1624	6.3850	156.6
19	0.03589	1288	8.0510	124.2
20	0.03196	1022	10.1500	98.5
21	0.02850		12.8000	78.1
22	0.02540		16.1400	62.0
23	0.02260		20.3600	49.1
24	0.02010		25.6700	39.0
25	0.01790		32.3700	30.9
26	0.01590		40.8100	24.5
27	0.01420		51.4700	19.4
28	0.01260		64.9000	15.4
29	0.01130		81.8300	12.2
30	0.01000		103.2000	9.7

RTDs and Thermocouples

The difference between RTDs and Thermocouples is that RTDs "Resistance Temperature Detectors" are basically resistors that change resistance at different temperatures. For instance, a 100 ohm RTD is called that because it will have 100 ohms of resistance at 0° C or 32° F. The resistance will change as the temperature changes.

A Thermocouple on the other hand is made up of two dissimilar metals that touch each other. The output is actually the reaction between the two metals, which is emitted in millivolts. As the temperature changes, the reactivity changes between the two metals, in turn changing the millivolt output. Different metals can be used in thermocouples and are classed as types. For example, a type "J" thermocouple is made up of Constantan and Iron, while a type "K" is made up of Chromel and Alumel.

RESISTANCE AT DEGREES F.							
FOR BURNS 100 OHM PLATINUM RESISTANCE BULB AT 32 DEG. F. (For 200 & 500 ohm bulbs multiply following resistance values by 2 and 5 respectively)							
Temp. F.	Ohms	Temp. F.	Ohms	Temp. F.	Ohms	Temp. F.	Ohms
-50	81.83	105	115.97	260	149.23	415	181.62
-45	82.94	110	117.05	265	150.29	420	182.65
-40	84.06	115	118.14	270	151.34	425	183.68
-35	85.17	120	119.22	275	152.4	430	184.71
-30	86.29	125	120.31	280	153.46	435	185.74
-25	87.4	130	121.39	285	154.51	440	186.76
-20	88.51	135	122.47	290	155.56	445	187.79
-15	89.62	140	123.55	295	156.62	450	188.81
-10	90.73	145	124.63	300	157.67	455	189.84
-5	91.83	150	125.71	305	158.72	460	190.86
0	92.94	155	126.79	310	159.77	465	191.88
5	94.05	160	127.87	315	160.82	470	192.91
10	95.15	165	128.94	320	161.87	475	193.93
15	96.25	170	130.02	325	162.92	480	194.95
20	97.36	175	131.09	330	163.96	485	195.96
25	98.46	180	132.17	335	165.01	490	196.98
30	99.56	185	133.24	340	166.05	495	198
35	100.66	190	134.31	345	167.1	500	199.02
40	101.76	195	135.38	350	168.14	520	203.07
45	102.86	200	136.46	355	169.18	540	207.11
50	103.95	205	137.52	360	170.23	560	211.14
55	105.05	210	138.59	365	171.27	580	215.15
60	106.16	215	139.66	370	172.31	600	219.15
65	107.24	220	140.73	375	173.34	650	229.08
70	108.34	225	141.79	380	174.38	700	238.91
75	109.43	230	142.86	385	175.42	750	248.66
80	110.52	235	143.92	390	176.45	800	258.32
85	111.61	240	144.98	395	177.49	850	267.89
90	112.7	245	146.05	400	178.52	900	277.37
95	113.79	250	147.11	405	179.56	950	286.76
100	114.88	255	148.17	410	180.59	1000	296.05

TEMPERATURE CONVERSION TABLE

Look up reading in middle column. If in degrees C., read F. equivelent in right hand colum
if in degrees F., read C. equivelent in left hand column.)

C		F	C		F	C		F
-46	-50	-58	6.7	44	111.2	33.9	93	199.4
-40	-40	-40	7.2	45	113	34.4	94	201.2
-34	-30	-22	7.8	46	114.8	35	95	203
-29	-20	-4	8.3	47	116.6	35.6	96	204.8
-23	-10	14	8.9	48	118.4	36.1	97	206.6
-17.8	0	32	9.4	49	120.2	36.7	98	208.4
-17.2	1	33.8	10	50	122	37.2	99	210.2
-16.7	2	35.6	10.6	51	123.8	37.8	100	212
-16.1	3	37.4	11.1	52	125.6	43	110	230
-15.6	4	39.2	11.7	53	127.4	49	120	248
-15	5	41	12.2	54	129.2	54	130	266
-14.4	6	42.8	12.8	55	131	60	140	284
-13.9	7	44.6	13.3	56	132.8	66	150	302
-13.3	8	46.4	13.9	57	134.6	71	160	320
-12.8	9	48.2	14.4	58	136.4	77	170	338
-12.2	10	50	15	59	138.2	82	180	356
-11.7	11	51.8	15.6	60	140	88	190	374
-11.1	12	53.6	16.1	61	141.8	93	200	392
-10.6	13	55.4	16.7	62	143.6	99	210	410
-10	14	57.2	17.2	63	145.4	100	212	413.6
-9.4	15	59	17.8	64	147.2	104	220	428
-8.9	16	60.8	18.3	65	149	110	230	446
-8.3	17	62.6	18.9	66	150.8	116	240	464
-7.8	18	64.4	19.4	67	152.6	121	250	482
-7.2	19	66.2	20	68	154.4	127	260	500
-6.7	20	68	20.6	69	156.2	132	270	518
-6.1	21	69.8	21.1	70	158	138	280	536
-5.6	22	71.6	21.7	71	159.8	143	290	554
-5	23	73.4	22.2	72	161.6	149	300	572
-4.4	24	75.2	22.8	73	163.4	154	310	590
-3.9	25	77	23.3	74	165.2	160	320	608
-3.3	26	78	23.9	75	167	166	330	626
-2.8	27	80.6	24.4	76	168.8	171	340	644
-2.2	28	82.4	25	77	170.6	177	350	662
-1.7	29	84.2	25.6	78	172.4	182	360	680
-1.1	30	86	26.1	79	174.2	188	370	698
-0.6	31	87.8	26.7	80	176	193	380	716
0	32	89.6	27.2	81	177.8	199	390	734
0.6	33	91.4	27.8	82	179.6	204	400	752
1.1	34	93.2	28.3	83	181.4	210	410	770
1.7	35	95	28.9	84	183.2	216	420	788
2.2	36	96.8	29.4	85	185	221	430	806
2.8	37	98.6	30	86	186.8	227	440	824
3.3	38	100.4	30.6	87	188.6	232	450	842
3.9	39	102.2	31.1	88	190.4	238	460	860
4.4	40	104	31.7	89	192.2	243	470	878
5	41	105.8	32.2	90	194	249	480	896
5.6	42	107.6	32.8	91	195.8	254	490	914
6.1	43	109.4	33.3	92	197.6	260	500	932

RANGES OF THERMOCOUPLES AND COLOR CODES					
THERMOCOUPLES	TYPE	DEGREES F.	COLOR	COLOR	EMF (MV)
Copper/Constantan	T	-300 to 750	+	-	-5.284 to 20.805
Iron/Constantan	J	-300 to 1600	+	-	-7.52 to 50.05
Chromel/Alumel	K	-300 to 2300	+	-	-5.51 to 51.05
Chromel/Constantan	E	32 to 1800	+	-	0 to 75.12
Platinum/Rhodium	R	32 to 3200	+	-	0 to 21.006
Platinum/Rhodium	S	32 to 3200	+	-	0 to 18.612
Nicrosil/Nisil	N	-450 to 2370	+	-	.-4.344 to 47.462
Platinum/Rhodium	B	32 to 3100	+	-	0 to 13.814
Tungsten/Rhenium	C	32 to 4200	+	-	
Nickel/Molybdenum	M	32 to 2550	+	-	
Chromel/Gold/Iron			+	-	
Tungsten/Rhenium	D	32 to 4200	+	-	
Tungsten/Rhenium	G	32 to 4200	+	-	
Copper/Copper/Nickel	U	-300 to 750	+	-	
Copper/Copper	V		+	-	
Tungsten/Rhenium	W	32 to 4200	+	-	

CONDUCTOR RESISTANCE	
MATERIAL	SPECIFIC RESISTANCE
SILVER	9.75
COPPER	10.55
ALUMINUM	17.3
NICKEL	53
IRON	61
LEAD	115
NICHROME	660

THERMOCOUPLE MILLIVOLT TABLE

Color Jacket	Type K	Type E	Type J	Type T
Color +	+	+	+	+
Color -	-	-	-	-
RANGE	-184 C.	-217 C.	-204 C.	-225 C.
Deg. C.	1373 C.	1001 C.	1200 C.	400 C.
-40	-1.5270	-2.2550	-2.4310	-1.8190
-30	-1.1560	-1.7090	-1.9610	-1.4750
-20	-0.7780	-1.1520	-1.4820	-1.1210
-10	-0.3920	-0.5820	-0.9950	-0.7570
0	0.0000	0.0000	0.0000	0.0000
10	0.3970	0.5910	0.5580	0.3910
20	0.7980	1.1920	1.0190	0.7900
30	1.2030	1.8010	1.5370	1.1960
40	1.6120	2.4200	2.0590	1.6120
50	2.0230	3.0480	2.5850	2.0360
60	2.4360	3.6850	3.1160	2.4680
70	2.8510	4.3300	3.6500	2.9090
80	3.2670	4.9850	4.1870	3.3580
90	3.6820	5.6480	4.7260	3.8140
100	4.0960	6.3190	5.2690	4.2790
110	4.5090	6.9980	5.8140	4.7500
120	4.9200	7.6850	6.3600	5.2280
130	5.3280	8.3790	6.9090	5.7140
140	5.7350	9.0810	7.4590	6.2060
150	6.1380	9.7890	8.0100	6.7040
160	6.5400	10.5030	8.5620	7.2090
170	6.9410	11.2240	9.1150	7.7200
180	7.4300	11.9510	9.6690	8.2370
190	7.7390	12.6840	10.2240	8.7590
200	8.1380	13.4210	10.7790	9.2880
210	8.5390	14.1640	11.3340	9.8220
220	8.9400	14.9120	11.8890	10.3620
230	9.3430	15.6640	12.4450	10.9070
240	9.7470	16.4200	13.0000	11.4580
250	10.1530	17.1810	13.5550	12.0130

THERMOCOUPLE MILLIVOLT TABLE at °F								
Jacket	Type B	Type E	Type J	Type K	Type N	Type R	Type S	Type T
Color +	+	+	+	+	+	+	+	+
Color -	-	-	-	-	-	-	-	-
-450		-9.830		-6.456	-4.344			-6.254
-440		-9.809		-6.447	-4.339			-6.240
-430		-9.775		-6.431	-4.330			-6.217
-420		-9.729		-6.409	-4.316			-6.187
-410		-9.672		-6.380	-4.299			-6.150
-400		-9.604		-6.344	-4.277			-6.105
-390		-9.526		-6.301	-4.251			-6.053
-380		-9.437		-6.251	-4.220			-5.995
-370		-9.338		-6.195	-4.185			-5.930
-360		-9.229		-6.133	-4.145			-5.860
-350		-9.112		-6.064	-4.102			-5.785
-340		-8.986	-8.030	-5.989	-4.054			-5.705
-330		-8.852	-7.915	-5.908	-4.001			-5.620
-320		-8.710	-7.791	-5.822	-3.945			-5.532
-310		-8.561	-7.659	-5.730	-3.884			-5.439
-300		-8.404	-7.519	-5.632	-3.820			-5.341
-290		-8.240	-7.372	-5.529	-3.752			-5.240
-280		-8.069	-7.218	-5.421	-3.679			-5.135
-270		-7.891	-7.057	-5.308	-3.603			-5.025
-260		-7.707	-6.890	-5.190	-3.524			-4.912
-250		-7.516	-6.716	-5.067	-3.441			-4.794
-240		-7.319	-6.536	-4.939	-3.354			-4.673
-230		-7.116	-6.350	-4.806	-3.264			-4.548
-220		-6.907	-6.159	-4.669	-3.170			-4.419
-210		-6.692	-5.962	-4.527	-3.074			-4.286
-200		-6.471	-5.760	-4.381	-2.974			-4.149
-190		-6.245	-5.553	-4.230	-2.871			-4.009
-180		-6.013	-5.341	-4.075	-2.765			-3.864
-170		-5.776	-5.124	-3.917	-2.656			-3.717
-160		-5.534	-4.903	-3.754	-2.544			-3.565

Deg.	Type B	Type E	Type J	Type K	Type N	Type R	Type S	Type T
-150		-5.287	-4.678	-3.587	-2.430			-3.410
-140		-5.034	-4.448	-3.417	-2.313			-3.251
-130		-4.777	-4.215	-3.242	-2.193			-3.089
-120		-4.515	-3.978	-3.065	-2.071			-2.923
-110		-4.248	-3.737	-2.883	-1.947			-2.753
-100		-3.976	-3.492	-2.699	-1.821			-2.581
-90		-3.700	-3.245	-2.511	-1.692			-2.405
-80		-3.419	-2.994	-2.320	-1.562			-2.225
-70		-3.134	-2.740	-2.126	-1.429			-2.042
-60		-2.845	-2.483	-1.929	-1.295			-1.856
-50		-2.552	-2.223	-1.729	-1.160	-0.210	-0.218	-1.667
-40		-2.254	-1.960	-1.527	-1.023	-0.188	-0.194	-1.475
-30		-1.953	-1.695	-1.322	-0.884	-0.165	-0.170	-1.279
-20		-1.648	-1.428	-1.114	-0.744	-0.141	-0.145	-1.081
-10		-1.339	-1.158	-0.904	-0.603	-0.116	-0.119	-0.879
0		-1.026	-0.885	-0.692	-0.461	-0.089	-0.092	-0.674
10		-0.709	-0.611	-0.478	-0.318	-0.063	-0.064	-0.467
20		-0.389	-0.334	-0.262	-0.174	-0.035	-0.035	-0.256
30		0.065	-0.056	-0.044	-0.029	-0.006	-0.006	-0.043
40	-0.001	0.262	0.224	0.176	0.115	0.024	0.024	0.173
50	-0.002	0.591	0.507	0.397	0.261	0.054	0.055	0.391
60	-0.002	0.924	0.791	0.619	0.407	0.086	0.087	0.611
70	-0.003	1.259	1.076	0.843	0.554	0.118	0.119	0.834
80	-0.002	1.597	1.363	1.068	0.703	0.150	0.152	1.060
90	-0.002	1.937	1.652	1.294	0.853	0.184	0.186	1.288
100	-0.001	2.281	1.942	1.520	1.003	0.218	0.221	1.518
110	0.000	2.627	2.233	1.748	1.156	0.253	0.256	1.752
120	0.002	2.977	2.526	1.977	1.309	0.289	0.291	1.988
130	0.004	3.329	2.820	2.206	1.463	0.326	0.328	2.226
140	0.006	3.683	3.115	2.436	1.619	0.363	0.365	2.467
150	0.009	4.041	3.411	2.666	1.775	0.400	0.402	2.711
160	0.012	4.401	3.708	2.896	1.933	0.439	0.440	2.958
170	0.015	4.764	4.006	3.127	2.092	0.478	0.478	3.206
180	0.019	5.130	4.305	3.358	2.253	0.517	0.517	3.458

Deg.	Type B	Type E	Type J	Type K	Type N	Type R	Type S	Type T
190	0.023	5.498	4.605	3.589	2.414	0.557	0.557	3.711
200	0.027	5.869	4.906	3.819	2.577	0.598	0.597	3.967
210	0.032	6.242	5.207	4.049	2.741	0.639	0.637	4.225
220	0.037	6.618	5.509	4.279	2.906	0.681	0.678	4.486
230	0.043	6.996	5.812	4.508	3.072	0.723	0.719	4.749
240	0.049	7.377	6.116	4.737	3.239	0.766	0.761	5.014
250	0.055	7.760	6.420	4.964	3.407	0.809	0.803	5.281
260	0.061	8.145	6.724	5.192	3.577	0.852	0.846	5.550
270	0.068	8.532	7.029	5.418	3.748	0.897	0.889	5.821
280	0.075	8.922	7.335	5.643	3.919	0.941	0.932	6.094
290	0.083	9.314	7.641	5.868	4.092	0.986	0.976	6.369
300	0.090	9.708	7.947	6.092	4.266	1.032	1.020	6.647
310	0.099	10.103	8.253	6.316	4.441	1.077	1.064	6.926
320	0.107	10.501	8.560	6.539	4.617	1.124	1.109	7.207
330	0.116	10.901	8.867	6.761	4.794	1.170	1.154	7.490
340	0.125	11.302	9.175	6.984	4.971	1.217	1.199	7.775
350	0.135	11.706	9.483	7.205	5.150	1.265	1.245	8.062
360	0.144	12.111	9.790	7.427	5.330	1.313	1.291	8.350
370	0.155	12.518	10.098	7.649	5.511	1.361	1.337	8.641
380	0.165	12.926	10.407	7.870	5.693	1.409	1.384	8.933
390	0.176	13.336	10.715	8.092	5.875	1.458	1.431	9.227
400	0.187	13.748	11.023	8.314	6.059	1.508	1.478	9.523
410	0.199	14.161	11.332	8.537	6.243	1.557	1.525	9.820
420	0.210	14.576	11.640	8.759	6.429	1.607	1.573	10.120
430	0.223	14.992	11.949	8.983	6.615	1.657	1.620	10.420
440	0.235	15.410	12.257	9.206	6.802	1.708	1.669	10.723
450	0.248	15.829	12.566	9.430	6.989	1.758	1.717	11.027
460	0.261	16.249	12.874	9.655	7.178	1.810	1.765	11.333
470	0.275	16.670	13.183	9.880	7.367	1.861	1.814	11.640
480	0.288	17.093	13.491	10.106	7.557	1.913	1.863	11.949
490	0.303	17.517	13.800	10.333	7.748	1.964	1.912	12.260
500	0.317	17.942	14.108	10.560	7.940	2.017	1.962	12.572
510	0.332	18.368	14.416	10.787	8.132	2.069	2.011	12.885
520	0.347	18.795	14.724	11.015	8.325	2.122	2.061	13.200

Deg.	Type B	Type E	Type J	Type K	Type N	Type R	Type S	Type T
530	0.362	19.223	15.032	11.243	8.519	2.175	2.111	13.516
540	0.378	19.653	15.340	11.472	8.713	2.228	2.161	13.834
550	0.394	20.083	15.648	11.702	8.909	2.282	2.211	14.153
560	0.410	20.514	15.956	11.931	9.104	2.335	2.262	14.474
570	0.427	20.947	16.264	12.161	9.301	2.389	2.313	14.795
580	0.444	21.380	16.571	12.392	9.498	2.443	2.363	15.118
590	0.462	21.814	16.879	12.623	9.695	2.498	2.414	15.443
600	0.479	22.248	17.186	12.854	9.894	2.552	2.465	15.769
610	0.497	22.684	17.493	13.085	10.092	2.607	2.517	16.096
620	0.515	23.120	17.800	13.317	10.292	2.662	2.568	16.424
630	0.534	23.558	18.107	13.549	10.492	2.718	2.620	16.753
640	0.553	23.996	18.414	13.781	10.692	2.773	2.672	17.084
650	0.572	24.434	18.721	14.013	10.893	2.829	2.723	17.416
660	0.592	24.873	19.027	14.246	11.095	2.885	2.775	17.750
670	0.612	25.313	19.334	14.479	11.297	2.941	2.828	18.084
680	0.632	25.754	19.640	14.712	11.499	2.997	2.880	18.420
690	0.652	26.195	19.947	14.945	11.703	3.053	2.932	18.757
700	0.673	26.637	20.253	15.178	11.906	3.110	2.985	19.095
710	0.694	27.079	20.559	15.412	12.110	3.167	3.037	19.434
720	0.716	27.522	20.866	15.646	12.315	3.224	3.090	19.774
730	0.737	27.966	21.172	15.880	12.520	3.281	3.143	20.116
740	0.759	28.409	21.478	16.114	12.725	3.338	3.196	20.458
750	0.782	28.854	21.785	16.349	12.931	3.396	3.249	20.801
760	0.804	29.299	22.091	16.583	13.137	3.453	3.302	
770	0.827	29.744	22.397	16.818	13.344	3.511	3.356	
780	0.851	30.190	22.704	17.053	13.551	3.569	3.409	
790	0.874	30.636	23.010	17.288	13.759	3.627	3.463	
800	0.898	31.082	23.317	17.523	13.967	3.686	3.516	
810	0.922	31.529	23.624	17.759	14.175	3.744	3.570	
820	0.947	31.976	23.931	17.994	14.384	3.803	3.624	
830	0.972	32.423	24.238	18.230	14.593	3.862	3.678	
840	0.997	32.871	24.546	18.466	14.802	3.921	3.732	
850	1.022	33.319	24.853	18.702	15.012	3.980	3.786	
860	1.048	33.767	25.161	18.938	15.222	4.039	3.840	

Deg.	Type B	Type E	Type J	Type K	Type N	Type R	Type S	Type T
870	1.074	34.215	25.469	19.174	15.432	4.099	3.895	
880	1.100	34.664	25.778	19.410	15.643	4.158	3.949	
890	1.127	35.113	26.087	19.646	15.854	4.218	4.004	
900	1.153	35.562	26.396	19.883	16.066	4.278	4.058	
910	1.181	36.011	26.705	20.120	16.278	4.338	4.113	
920	1.208	36.460	27.016	20.356	16.490	4.398	4.168	
930	1.236	36.909	27.326	20.593	16.702	4.458	4.223	
940	1.264	37.358	27.637	20.830	16.915	4.519	4.278	
950	1.292	37.808	27.949	21.066	17.127	4.580	4.333	
960	1.321	38.257	28.261	21.303	17.341	4.640	4.388	
970	1.350	38.707	28.573	21.540	17.554	4.701	4.443	
980	1.379	39.157	28.887	21.777	17.768	4.762	4.498	
990	1.409	39.606	29.201	22.014	17.982	4.824	4.554	
1000	1.438	40.056	29.515	22.251	18.196	4.885	4.609	
1010	1.468	40.505	29.831	22.488	18.410	4.947	4.665	
1020	1.499	40.955	30.147	22.725	18.625	5.008	4.721	
1030	1.529	41.404	30.464	22.961	18.840	5.070	4.776	
1040	1.560	41.853	30.782	23.198	19.055	5.132	4.832	
1050	1.591	42.303	31.100	23.435	19.270	5.194	4.888	
1060	1.623	42.752	31.420	23.672	19.486	5.256	4.944	
1070	1.655	43.201	31.740	23.908	19.701	5.319	5.000	
1080	1.687	43.650	32.061	24.145	19.917	5.381	5.057	
1090	1.719	44.098	32.384	24.382	20.133	5.444	5.113	
1100	1.752	44.547	32.707	24.618	20.350	5.507	5.169	
1110	1.785	44.995	33.031	24.854	20.566	5.570	5.226	
1120	1.818	45.443	33.356	25.091	20.783	5.633	5.283	
1130	1.851	45.891	33.683	25.327	20.999	5.696	5.339	
1140	1.885	46.339	34.010	25.563	21.216	5.759	5.396	
1150	1.919	46.786	34.339	25.799	21.433	5.823	5.453	
1160	1.953	47.234	34.668	26.034	21.650	5.886	5.510	
1170	1.988	47.681	34.999	26.270	21.868	5.950	5.567	
1180	1.022	48.127	35.331	26.505	22.085	6.014	5.625	
1190	2.058	48.574	35.664	26.740	22.302	6.078	5.682	
1200	2.093	49.020	35.999	26.975	22.520	6.143	5.740	

Deg.	Type B	Type E	Type J	Type K	Type N	Type R	Type S	Type T
1210	2.128	49.466	36.334	27.210	22.738	6.207	5.797	
1220	2.164	49.911	36.671	27.445	22.956	6.272	5.855	
1230	2.201	50.357	37.009	27.679	23.173	6.336	5.913	
1240	2.237	50.802	37.348	27.914	23.391	6.401	5.971	
1250	2.274	51.246	37.688	28.148	23.609	6.466	6.029	
1260	2.311	51.691	38.030	28.382	23.828	6.532	6.087	
1270	2.348	52.135	38.372	28.615	24.046	6.597	6.146	
1280	2.385	52.578	38.716	28.849	24.264	6.662	6.204	
1290	2.423	53.022	39.061	29.082	24.482	6.728	6.263	
1300	2.461	53.465	39.407	29.315	24.700	6.794	6.321	
1310	2.499	53.907	39.754	29.547	24.919	6.860	6.380	
1320	2.538	54.349	40.103	29.780	25.137	6.926	6.439	
1330	2.576	54.791	40.452	30.012	25.356	6.992	6.498	
1340	2.615	55.233	40.802	30.244	25.574	7.059	6.557	
1350	2.655	55.674	41.154	30.475	25.792	7.125	6.616	
1360	2.694	56.115	41.506	30.706	26.011	7.192	6.675	
1370	2.734	56.555	41.859	30.937	26.229	7.259	6.734	
1380	2.774	56.995	42.212	31.168	26.448	7.326	6.794	
1390	2.814	57.434	42.567	31.399	26.666	7.393	6.853	
1400	2.855	57.873	42.919	31.629	26.885	7.460	6.913	
1410	2.896	58.312	43.274	31.859	27.103	7.527	6.972	
1420	2.937	58.750	43.631	32.068	27.321	7.595	7.032	
1430	2.978	59.188	43.988	32.317	27.540	7.663	7.092	
1440	3.019	59.626	44.346	32.546	27.758	7.731	7.152	
1450	3.061	60.063	44.705	32.775	27.976	7.799	7.212	
1460	3.103	60.499	45.064	33.003	28.194	7.867	7.272	
1470	3.146	60.935	45.423	33.231	28.413	7.935	7.333	
1480	3.188	61.371	45.782	33.459	28.631	8.004	7.393	
1490	3.231	61.806	46.141	33.686	28.849	8.072	7.454	
1500	3.274	62.240	46.500	33.913	29.067	8.141	7.514	
1510	3.317	62.675	46.858	34.140	29.285	8.210	7.575	
1520	3.361	63.108	47.216	34.366	29.502	8.279	7.636	
1530	3.404	63.542	47.574	34.593	29.720	8.348	7.697	
1540	3.448	63.974	47.931	34.818	29.938	8.417	7.758	

Deg.	Type B	Type E	Type J	Type K	Type N	Type R	Type S	Type T
1550	3.492	64.406	48.288	35.044	30.156	8.487	7.819	
1560	3.537	64.838	48.644	35.269	30.373	8.556	7.880	
1570	3.581	65.269	48.999	35.494	30.591	8.626	7.942	
1580	3.626	65.700	49.353	35.718	30.808	8.696	8.003	
1590	3.672	66.130	49.707	35.942	31.025	8.766	8.065	
1600	3.717	66.559	50.060	36.166	31.242	8.836	8.126	
1610	3.762	66.988	50.411	36.390	31.460	8.907	8.188	
1620	3.808	67.416	50.762	36.613	31.677	8.977	8.250	
1630	3.854	67.844	51.112	36.836	31.893	9.048	8.312	
1640	3.901	68.271	51.460	37.058	32.110	9.118	8.374	
1650	3.947	68.698	51.808	37.280	32.327	9.189	8.436	
1660	3.994	69.124	52.154	37.502	32.543	9.260	8.498	
1670	4.041	69.549	52.500	37.724	32.760	9.331	8.560	
1680	4.088	69.974	52.844	37.945	32.976	9.403	8.623	
1690	4.136	70.398	53.188	38.166	33.192	9.474	8.685	
1700	4.183	70.821	53.530	38.387	33.408	9.546	8.748	
1710	4.231	71.244	53.871	38.607	33.624	9.617	8.811	
1720	4.279	71.667	54.211	38.827	33.840	9.689	8.874	
1730	4.327	72.088	54.550	39.046	34.056	9.761	8.937	
1740	4.376	72.509	54.888	39.266	34.271	9.833	9.000	
1750	4.425	72.930	55.225	39.485	34.487	9.906	9.063	
1760	4.474	73.350	55.561	39.703	34.702	9.978	9.126	
1770	4.523	73.769	55.896	39.922	34.917	10.050	9.190	
1780	4.572	74.188	56.230	40.140	35.132	10.123	9.253	
1790	4.622	74.606	56.564	40.358	35.347	10.196	9.317	
1800	4.672	75.024	56.896	40.575	35.562	10.269	9.380	
1810	4.722	75.441	57.227	40.792	35.777	10.342	9.444	
1820	4.772	75.858	57.558	41.009	35.991	10.415	9.508	
1830	4.823	76.274	57.888	41.225	36.205	10.488	9.572	
1840	4.873		58.217	41.442	36.419	10.562	9.636	
1850	4.924		58.545	41.657	36.633	10.636	9.700	
1860	4.975		58.872	41.873	36.847	10.709	9.764	
1870	5.027		59.199	42.088	37.061	10.783	9.829	
1880	5.078		59.526	42.303	37.274	10.857	9.893	

Deg.	Type B	Type E	Type J	Type K	Type N	Type R	Type S	Type T
1890	5.130		59.851	42.518	37.488	10.931	9.958	
1900	5.182		60.177	42.732	37.701	11.006	10.023	
1910	5.234		60.501	42.946	37.914	11.080	10.087	
1920	5.286		60.826	43.159	38.127	11.155	10.152	
1930	5.339		61.149	43.373	38.340	11.229	10.217	
1940	5.391		61.473	43.585	38.552	11.304	10.282	
1950	5.444		61.796	43.798	38.764	11.379	10.348	
1960	5.497		62.118	44.010	38.976	11.454	10.413	
1970	5.551		62.441	44.222	39.188	11.529	10.478	
1980	5.604		62.763	44.434	39.400	11.605	10.544	
1990	5.658		63.085	44.645	39.612	11.680	10.609	
2000	5.712		63.406	44.856	39.823	11.756	10.675	
2010	5.766		63.728	45.066	40.034	11.831	10.740	
2020	5.820		64.049	45.276	40.245	11.907	10.806	
2030	5.875		64.370	45.486	40.456	11.983	10.872	
2040	5.930		64.691	45.695	40.666	12.059	10.938	
2050	5.984		65.012	45.904	40.877	12.135	11.004	
2060	6.039		65.333	46.113	41.087	12.211	11.070	
2070	6.095		65.654	46.321	41.297	12.287	11.136	
2080	6.150		65.974	46.529	41.506	12.363	11.202	
2090	6.206		66.295	46.737	41.716	12.440	11.268	
2100	6.262		66.615	46.944	41.925	12.516	11.335	
2110	6.316		66.935	47.150	42.134	12.593	11.401	
2120	6.374		67.255	47.356	42.342	12.669	11.467	
2130	6.430		67.575	47.562	42.550	12.746	11.534	
2140	6.487		67.895	47.767	42.759	12.823	11.600	
2150	6.543		68.214	47.972	42.966	12.900	11.667	
2160	6.600		68.534	48.177	43.174	12.977	11.734	
2170	6.657		68.853	48.381	43.381	13.054	11.800	
2180	6.714		69.171	48.584	43.588	13.131	11.867	
2190	6.772		69.490	48.787	43.795	13.208	11.934	
2200	6.829			48.990	44.001	13.286	12.001	
2210	6.887			49.192	44.207	13.363	12.067	
2220	6.945			49.394	44.413	13.440	12.134	

Deg.	Type B	Type E	Type J	Type K	Type N	Type R	Type S	Type T
2230	7.003			49.595	44.619	13.518	12.201	
2240	7.061			49.796	44.824	13.595	12.268	
2250	7.120			49.996	45.029	13.673	12.335	
2260	7.178			50.196	45.233	13.751	12.402	
2270	7.237			50.395	45.437	13.828	12.469	
2280	7.296			50.594	45.641	13.906	12.536	
2290	7.355			50.792	45.845	13.984	12.604	
2300	7.414			50.990	46.048	14.062	12.671	
2310	7.473			51.187	46.251	14.140	12.738	
2320	7.533			51.384	46.453	14.218	12.805	
2330	7.592			51.580	46.656	14.296	12.872	
2340	7.652			51.776	46.858	14.374	12.940	
2350	7.712			51.971	47.059	14.452	13.007	
2360	7.772			52.165	47.261	14.530	13.074	
2370	7.833			52.360	47.462	14.608	13.142	
2380	7.893			52.553		14.686	13.209	
2390	7.953			52.747		14.765	13.276	
2400	8.014			52.939		14.843	13.344	
2410	8.075			53.132		14.921	13.411	
2420	8.136			53.324		15.000	13.478	
2430	8.197			53.515		15.078	13.546	
2440	8.258			53.706		15.156	13.613	
2450	8.319			53.897		15.235	13.681	
2460	8.381			54.087		15.313	13.748	
2470	8.442			54.277		15.391	13.815	
2480	8.504			54.466		15.470	13.883	
2490	8.566			54.656		15.548	13.950	
2500	8.628			54.845		15.627	14.018	
2510	8.690					15.705	14.085	
2520	8.752					15.784	14.152	
2530	8.814					15.862	14.220	
2540	8.877					15.941	14.287	
2550	8.939					16.019	14.354	
2560	9.002					16.097	14.422	

Deg.	Type B	Type E	Type J	Type K	Type N	Type R	Type S	Type T
2570	9.065					16.176	14.489	
2580	9.128					16.254	14.556	
2590	9.191					16.333	14.624	
2600	9.254					16.411	14.691	
2610	9.317					16.490	14.758	
2620	9.380					16.568	14.826	
2630	9.443					16.646	14.893	
2640	9.507					16.725	14.960	
2650	9.570					16.803	15.027	
2660	9.634					16.882	15.094	
2670	9.697					16.960	15.161	
2680	9.761					17.038	15.228	
2690	9.825					17.116	15.295	
2700	9.889					17.195	15.362	
2710	9.953					17.273	15.429	
2720	10.017					17.351	15.496	
2730	10.081					17.429	15.563	
2740	10.145					17.507	15.630	
3750	10.210					17.585	15.697	
2760	10.274					17.663	15.763	
2770	10.338					17.741	15.830	
2780	10.403					17.819	15.897	
2790	10.467					17.897	15.963	
2800	10.532					17.975	16.030	
2810	10.596					18.053	16.096	
2820	10.661					18.130	16.163	
2830	10.726					18.208	16.229	
2840	10.790					18.286	16.296	
2850	10.855					18.363	16.362	
2860	10.920					18.441	16.428	
2870	10.985					18.518	16.494	
2880	11.050					18.595	16.560	
2890	11.115					18.673	16.626	
2900	11.179					18.750	16.692	

Deg.	Type B	Type E	Type J	Type K	Type N	Type R	Type S	Type T
2910	11.244					18.827	16.758	
2920	11.309					18.904	16.824	
2930	11.374					18.981	16.890	
2940	11.439					19.058	16.955	
2950	11.504					19.135	17.021	
2960	11.569					19.211	17.086	
2970	11.634					19.288	17.152	
2980	11.699					19.365	17.217	
2990	11.764					19.441	17.282	
3000	11.829					19.518	17.347	
3010	11.894					19.594	17.412	
3020	11.959					19.670	17.477	
3030	12.024					19.746	17.542	
3040	12.089					19.822	17.607	
3050	12.154					19.898	17.672	
3060	12.219					19.974	17.736	
3070	12.284					20.050	17.801	
3080	12.349					20.125	17.865	
3090	12.413					20.200	17.929	
3100	12.478					20.275	17.993	
3110	12.543					20.350	18.056	
3120	12.608					20.424	18.119	
3130	12.672					20.498	18.182	
3140	12.737					20.572	18.245	
3150	12.801					20.645	18.307	
3160	12.866					20.718	18.369	
3170	12.930					20.791	18.431	
3180	12.995					20.863	18.492	
3190	13.059					20.935	18.552	
3200	13.124					21.006	18.612	
3210	13.188					21.077	18.672	
3220	13.252							
3230	13.316							
3240	13.380							
3250	13.444							
3260	13.508							
3270	13.572							
3280	13.635							
3290	13.699							
3300	13.763							

THERMOMETRIC FIXED POINTS		
FIXED POINTS	MELTING POINTS IN DEGREES	
	DEGREES C.	DEGREES F.
BOILING POINT OF OXYGEN	-183	-297.3
SUBLIMATION POINTS OF CARBON DIOXIDE	-78.4	-109.2
FREEZING POINT OF MERCURY	-38.86	-37.95
FREEZING POINT OF WATER	0	32
TRIPLE POINT OF WATER	0.01	32
FREEZING POINT OF PHOSPHORUS	44	111
BOILING POINT OF WATER	100	212
TRIPLE POINT OF BENZOIC ACID	122.4	252.3
BOILING POINT OF NAPTHTHALENE	218	424.4
FREEZING POINT OF TIN	231.9	449.4
BOILING POINT OF BENZOPHENONE	305.9	582.6
FREEZING POINT OF CADMIUM	321.1	610
FREEZING POINT OF LEAD	327.5	621.5
FREEZING POINT OF ZINC	419.6	787.2
BOILING POINT OF SULFER	444.7	832.4
FREEZING POINT OF ANTIMONY	630.7	1167.3
FREEZING POINT OF MAGNESIUM	650	1200
FREEZING POINT OF PLUTONIUM	640	1180
FREEZING POINT OF ALUMINUM	660.4	1220.7
FREEZING POINT OF ADMIRALTY BRASS	900	1650
FREEZING POINT OF YELLOW BRASS	930	1710
FREEZING POINT OF SILVER	961.9	1763.5
FREEZING POINT OF RED BRASS	1000	1832
FREEZING POINT OF GOLD	1064.4	1948
FREEZING POINT OF COPPER	1084.5	1984.1
FREEZING POINT OF GREY CAST IRON	1127	2060
FREEZING POINT OF URANIUM	1132	2070
FREEZING POINT OF DUCTILE IRON	1149	2100
FREEZING POINT OF BERYLLIUM	1285	2345
FREEZING POINT OF SILICON	1411	2572
FREEZING POINT OF CARBON STEEL	1425	2600
FREEZING POINT OF NICKEL	1453	2647
FREEZING POINT OF WROUGHT IRON	1482	2700
FREEZING POINT OF COBALT	1495	2723
FREEZING POINT OF STAINLESS STEEL	1510	2750
FREEZING POINT OF PALLADIUM	1554	2829
FREEZING POINT OF TITANIUM	1670	3040
FREEZING POINT OF PLATINUM	1772	3222
FREEZING POINT OF ZIRCONIUM	1854	3369
FREEZING POINT OF CHROMIUM	1860	3380
FREEZING POINT OF VANADIUM	1900	3450
FREEZING POINT OF IRIDIUM	2450	4440
FREEZING POINT OF MOLYBDENUM	2620	4750
FREEZING POINT OF TUNGSTEN	3400	6150

SPECIFIC GRAVITY OF LIQUIDS			
LIQUID	SP. GRAVITY	DEGREES C.	DEGREES F.
ACETIC ACID	1.052	25	77
ACETONE	0.787	25	77
ACETYLENE	0.38	21.1	70
ALCOHOL, ETHYL	0.787	25	77
ALCOHOL, METHYL	0.792	25	77
ALCOHOL, PROPYL	0.802	25	77
AMMONIA "AQUA"	0.826	25	77
ALUMINUM CHLORIDE 10%	1.073	20	68
ALUMINUM CHLORIDE 20%	1.154	20	68
ALUMINUM CHLORIDE 40%	1.342	20	68
AMMONIUM HYDROXIDE 10%	0.96	15	59
AMMONIUM HYDROXIDE 20%	0.925	15	59
AMMONIUM HYDROXIDE 30%	0.895	15	59
BENZENE	0.876	25	77
CHLORINE	1.56	-33.6	-28.5
ETHYLENE GLYCOL	1.07	15.6	60
GASOLINE	0.751	15.6	60
GLYCERIN	1.26	0	32
HYDROCHLORIC ACID 10%	1.047	20	68
HYDROCHLORIC ACID 20%	1.098	20	68
HYDROCHLORIC ACID 30%	1.149	20	68
KEROSENE	0.82	15.6	60
MERCURY	13.55	15.6	60
NITRIC ACID 10%	1.054	20	68
NITRIC ACID 30%	1.18	20	68
NITRIC ACID 50%	1.31	20	68
OIL, LINSEED	0.942	15	59
PHOSPHORIC ACID	1.87	20	68
POTASSIUM HYDROXIDE 10%	1.092	15	59
POTASSIUM HYDROXIDE 30%	1.291	15	59
POTASSIUM HYDROXIDE 50%	1.514	15	59
SILICONE (DC-200)	0.92	15.6	60
SODIUM CHLORIDE 10%	1.071	20	68
SODIUM CHLORIDE 20%	1.148	20	68
SODIUM HYDROXIDE 10%	1.109	20	68
SODIUM HYDROXIDE 30%	1.328	20	68
SODIUM HYDROXIDE 50%	1.525	20	68
SULFUR	1.78	572	1061.6
SULFURIC ACID 10%	1.066	20	68
SULFURIC ACID 20%	1.139	20	68
SULFURIC ACID 30%	1.218	20	68
SULFURIC ACID 40%	1.303	20	68
TURPENTINE	0.873	15.6	60
WATER	1	15.6	60
WATER, SEA	1.025	15	59

CONVERT "H2O TO PSI OR "Mercury					
"H2O @60 F.	PSI	"Hg @32 F.	"H2O @60 F.	PSI	"Hg @32 F.
1	0.03609	0.07348	29	1.04661	2.13092
2	0.07218	0.14696	30	1.0827	2.2044
3	0.10827	0.22044	31	1.11879	2.27788
4	0.14436	0.29392	32	1.15488	2.35136
5	0.18045	0.3674	33	1.19097	2.42464
6	0.21654	0.44088	34	1.22706	2.49832
7	0.25262	0.51436	35	1.26315	2.5718
8	0.28872	0.58784	36	1.29904	2.64528
9	0.32481	0.66132	37	1.33532	2.71876
10	0.3609	0.7348	38	1.37142	2.79224
11	0.39699	0.80838	39	1.40751	2.86572
12	0.43308	0.88196	40	1.4436	2.9392
13	0.46817	0.95524	41	1.47969	3.01268
14	0.50526	1.02872	42	1.51578	3.08616
15	0.54135	1.1022	43	1.55187	3.15964
16	0.57744	1.17568	44	1.58796	3.23312
17	0.61352	1.24916	45	1.62405	3.3066
18	0.64962	1.32264	46	1.66014	3.38008
19	0.68571	1.39612	47	1.69622	3.45356
20	0.7218	1.4696	48	1.73232	3.52704
21	0.75789	1.54308	49	1.76841	3.60052
22	0.79398	1.61656	50	1.8045	3.674
23	0.83007	1.69004	51	1.84059	3.74748
24	0.86616	1.76352	52	1.87668	3.82096
25	0.90225	1.837	53	1.91277	3.89444
26	0.93834	1.91048	54	1.94886	3.96792
27	0.97442	1.98396	55	1.98495	4.0414
28	1.01052	2.05744	56	2.02104	4.1148

"H2O @60 F.	PSI	"Hg @32 F.	"H2O @60 F.	PSI	"Hg @32 F.
57	2.05712	4.18836	84	3.03156	6.17232
58	2.0932	4.26184	85	3.06765	6.2458
59	2.12931	4.33532	86	3.13074	6.31828
60	2.1654	4.4088	87	3.13982	6.39276
61	2.20149	4.48228	88	3.17592	6.46624
62	2.23758	4.55576	89	3.21201	6.53972
63	2.27367	4.62924	90	3.2481	6.6132
64	2.30976	4.70272	91	3.28419	6.68668
65	2.34585	4.7752	92	3.32026	6.76016
66	2.38194	4.84968	93	3.35637	6.83364
67	2.41802	4.92316	94	3.39246	6.90712
68	2.45412	4.99664	95	3.42855	6.9806
69	2.49021	5.06412	96	3.46464	7.05408
70	2.5263	5.1436	97	3.50072	7.12756
71	2.56239	5.21708	98	3.53682	7.20104
72	2.59848	5.29056	99	3.57291	7.27452
73	2.63457	5.36401	100	3.609	7.348
74	2.67066	5.43752	200	7.218	14.696
75	2.70675	5.51	300	10.827	22.044
76	2.74284	5.58448	400	14.436	29.392
77	2.77892	5.65796	500	18.045	36.74
78	2.81502	5.73144	600	21.654	44.088
79	2.85111	5.80492	700	25.263	51.436
80	2.8872	5.8784	800	28.872	58.784
81	2.92329	5.95188	900	32.481	66.132
82	2.95938	6.02536	1000	36.09	73.48
83	2.99547	6.09884			

HANDY CONVERSIONS
1 CUBIC INCH = 16.3871 CUBIC CENTIMETERS
1 CUBIC FOOT = 0.0283168 CUBIC METERS
1 CUBIC FOOT = 1728 CUBIC INCHES
1 CUBIC YARD = 0.764555 CUBIC METERS
1 QUART = 0.946353 LITERS
1 INCH = 25.4 MILLIMETERS
1 INCH =2.54 CENTIMETERS
1 FOOT = 0.3048 METERS
1 YARD = 0.9144 METERS
1 NAUTICAL MILE = 1.1507797 MILES
1 MILE = 0.868976 NAUTICAL MILES
1 MILE = 1.60934 KILOMETERS
1 MILE = 1609.34 METERS
1 MILE = 1760 YARDS
1 MILE = 5280 FEET
1 SQUARE INCH = 6.4516 SQUARE CENTIMETERS
1 SQUARE INCH = 0.00064516 SQUARE METERS
1SQUARE INCH = 0.0069444 SQUARE FEET
1 SQUARE FOOT = 929.0304 SQUARE CENTIMETERS
1 SQUARE FOOT = 0.0929030 SQUARE METERS
1 SQUARE FOOT = 144 SQUARE INCHES
1 SQUARE YARD = 0.8361273 SQUARE METERS
1 SQUARE YARD = 8.3613 HECTARES
1 SQUARE MILE = 258.999 HECTARE
1 SQUARE MILE = 640 ACRE
1 ACRE = 0.404686 HECTARES
1 ACRE = 10 SQUARE CHAINS
1 HECTARE = 10,000 SQUARE METERS
1 CHAIN = 66 FEET
1 CHAIN = 22 YARDS
1 CHAIN = 20.1168 METERS
1 CHAIN = 4 RODS
1 CHAIN = 100 LINKS
1 ROD = 5.5 YARDS
1 ROD = 16.5 FEET
1 ROD = 198 INCHES
1 FATHOM = 72 INCHES
1 FATHOM = 6 FEET
1 FATHOM = 2 YARDS
1 FURLONG = 1/8 MILE
1 FURLONG = 10 CHAINS
1 FURLONG = 40 RODS
1 FURLONG = 660 FEET
1 FURLONG = 220 YARDS

1 STATUTE MILE = 80 CHAINS
1 GALLON = 0.00378541 CUBIC METERS
1 OUNCE = 28.3495 GRAMS
1 POUND = 0.4533592 KILOGRAMS
1 MILLIMETER = 0.0393701 INCHES
1 METER = 3.28084 FEET
1 METER = 1.09361 YARDS
1 KILOMETER = 0.621371 MILES
1 SQUARE METER = 10.7639 SQUARE FEET
1 SQUARE METER = 1.19599 SQUARE YARDS
1 CUBIC METER = 35.3147 CUBIC FEET
1 CUBIC METER = 1.30795 CUBIC YARDS
1 CUBIC METER = 264.172 GALLONS
1 SQUARE CENTIMETER = 0.1555 SQUARE INCHES
1 CUBIC CENTIMETER = 0.0610237 CUBIC INCHES
1 GRAM = 0.0352740 OUNCES
1 KILOGRAM = 2.20462 POUNDS
1 MIL = 1/1000 INCH = 0.001 INCHES
1 KILOWATT = 1.34102 HORSEPOWER
1 KILOWATT = 1000 WATTS
1 KILOWATT = 3.412 HEAT UNITS PER HOUR
1 KILOWATT = 56.9 HEAT UNITS PER MINUTE
1 KILOWATT = 0.948 HEAT UNITS PER SECOND
1 HEAT UNIT = 1.055 WATTS PER SECOND
1 HEAT UNIT = 0.000393 HORSEPOWER = HR
1 BTU = 3413 WATTS
1 WATT = .001 KILOWATTS
1 WATT = 0.001341 HORSEPOWER
1 HORSEPOWER = 745.71215 WATTS
1 AMP = 100 WATTS
1 PSI = 27.68 "H20
1 "H2O = 0.036 PSI
1 PSI = 2.02 "Hg
1 ATMOSPHERE = 14.7 PSIA
1 OUNCE = 1.73 "H2O
1 GALLON = 4 QUARTS
1 QUART = 2 PINTS
1 PINT = 2 CUPS
1 CUP = 8 US OZ.
1 CUP = 16 TABLE SPOONS
1 CUP = 48 TEA SPOONS
1 TABLE SPOON = 3 TEA SPOONS
1 US OZ. = 6 TSP
1 US OZ. = 2 TBSP

FRACTIONS AND DECIMAL EQUIVALENTS

Fraction	Decimal	mm	Fraction	Decimal	mm
1/64	0.01560	0.3969	**33/64**	0.51560	13.0969
1/32	0.03125	0.7938	**17/32**	0.53125	13.4938
3/64	0.04690	1.1906	**35/64**	0.54690	13.8906
1/16	0.06250	1.5875	**9/16**	0.56250	14.2875
5/64	0.07810	1.9874	**37/64**	0.57810	14.6844
3/32	0.09370	2.3813	**19/32**	0.59375	15.0813
7/64	0.10940	2.7781	**39/64**	0.60940	15.4781
1/8	0.12500	3.1750	**5/8**	0.62500	15.8750
9/64	0.14060	3.5719	**41/64**	0.64060	16.2719
5/32	0.15625	3.9688	**21/32**	0.65625	16.6688
11/64	0.17190	4.3656	**43/64**	0.67190	17.0656
3/16	0.18750	4.7625	**11/16**	0.68750	17.4625
13/64	0.20310	5.1594	**45/64**	0.70310	17.8594
7/32	0.21875	5.5563	**23/32**	0.71875	18.2563
15/64	0.23440	5.9531	**47/64**	0.73440	18.6531
1/4	0.25000	6.3500	**3/4**	0.75000	19.0500
17/64	0.26560	6.7469	**49/64**	0.76560	19.4469
9/32	0.28125	7.1438	**25/32**	0.78125	19.8438
19/64	0.29690	7.5406	**51/64**	0.79690	20.2406
5/16	0.31250	7.9375	**13/16**	0.81250	20.6375
21/64	0.32810	8.3344	**53/64**	0.82810	21.0344
11/32	0.34375	8.7313	**27/32**	0.84375	21.4313
23/64	0.35940	9.1281	**55/64**	0.85940	21.8281
3/8	0.37500	9.5250	**7/8**	0.87500	22.2250
25/64	0.39060	9.9219	**57/64**	0.89060	22.6219
13/32	0.40625	10.3188	**29/32**	0.90625	23.0188
27/64	0.42190	10.7156	**59/64**	0.92190	23.4156
7/16	0.43750	11.1125	**15/16**	0.93750	23.8125
29/64	0.45310	11.5094	**61/64**	0.95310	24.2094
15/32	0.46875	11.9063	**31/32**	0.96875	24.6063
31/64	0.48440	12.3031	**63/64**	0.98440	25.0031
1/2	0.50000	12.7000	**1**	1.00000	25.4000

TAP AND CLEARANCE DRILL SIZES FOR SCREWS									
SCREW SIZE				TAPPING DRILLS		CLEARANCE DRILLS			
						CLOSE FIT		FREE FIT	
SCREW SIZE	THREADS/In.	SERIES	DIAMETER	DRILL	DECIMAL	DRILL	DECIMAL	DRILL	DECIMAL
000	120		0.0340	71	0.0260	65	0.0350	62	0.0380
00	90		0.0470	65	0.0350	3/64"	0.0469	55	0.0520
0	80		0.0600	3/64"	0.0469	52	0.0635	50	0.0700
1	56	NS	0.0730	54	0.0550	48	0.0760	46	0.0810
1	64	NC	0.0730	53	0.0595	48	0.0760	46	0.0810
1	72	NF	0.0730	53	0.0595	48	0.0760	46	0.0810
2	56	NC	0.0860	50	0.0700	43	0.0890	41	0.0960
2	64	NF	0.0860	50	0.0700	43	0.0890	41	0.0960
3	48	NC	0.0990	47	0.0785	37	0.1040	35	0.1100
3	56	NF	0.0990	45	0.0785	37	0.1040	35	0.1100
4	36	NS	0.1120	44	0.0860	32	0.1160	30	0.1285
4	40	NC	0.1120	43	0.0890	32	0.1160	30	0.1285
4	48	NF	0.1120	42	0.0935	32	0.1160	30	0.1285
5	40	NC	0.1250	38	0.1015	30	0.1285	29	0.1360
5	44	NF	0.1250	37	0.1040	30	0.1285	29	0.1360
6	32	NC	0.1380	36	0.1065	27	0.1440	25	0.1495
6	36	NS	0.1380	34	0.1110	27	0.1440	25	0.1495
6	40	NF	0.1380	33	0.1130	27	0.1440	25	0.1495
8	32	NC	0.1640	29	0.1360	18	0.1695	16	0.1170
8	36	NS	0.1640	29	0.1360	18	0.1695	16	0.1170
8	40	NF	0.1640	28	0.1405	18	0.1695	16	0.1170
10	24	NC	0.1900	25	0.1495	9	0.1960	7	0.2010
10	30	NS	0.1900	22	0.1570	9	0.1960	7	0.2010
10	32	NF	0.1900	21	0.1590	9	0.1960	7	0.2010
12	24	NC	0.2160	16	0.1770	2	0.2210	1	0.2280
12	28	NF	0.2160	14	0.1820	2	0.2210	1	0.2280

DRILL & TAP SIZES for BOLTS

TAP SIZE	DECIMAL	DRILL SIZE	TAP SIZE	DECIMAL	DRILL SIZE
1/4-20	.2010	7	3/4-10	.6562	21/32
1/4-28	.213	3	3/4-16	.6875	11/16
5/16-18	.257	F	7/8-9	.7656	49/64
5/16-24	.272	I	7/8-14	.8125	13/16
3/8-16	.3125	5/16	1-8	.875	7/8
3/8-24	.332	Q	1-12	.9219	59/64
7/16-14	.368	U	1 1/8-7	.9844	63/64
7/16-20	.3906	25/64	1 1/8-12	1.0469	1 3/64
1/2-13	.4219	27/64	1 1/4-7	1.1094	1 7/64
1/2-20	.4531	29/64	1 1/4-12	1.1719	1 11/64
9/16-12	.4844	31/64	1 3/8-6	1.1288	1 7/32
9/16-18	.5156	33/64	1 3/8-12	1.2969	1 19/64
5/8-11	.5312	17/32	1 1/2-6	1.3438	1 11/32
5/8-18	.5781	37/64	1 1/2-12	1.4219	1 27/64

for PIPE

TAP SIZE	DRILL SIZE
1/8-27	11/32
1/4-18	7/16
3/8-18	37/64
1/2-14	23/32
3/4-14	59/64
1-11 1/2	1 5/32
1 1/4-11 1/2	1 1/2
1 1/2-11 1/2	1 3/4
2-11 1/2	2 7/32
2 1/2-8	2 21/32
3-8	3 1/4
3 1/2-8	3 3/4
4-8	4 1/4

AREA AND VOLUME

	AREA =	VOLUME =
CUBE	Length X Height	Length X Width X Height
RECTANGLE	Length X Height	Length X Width X Height
TRIANGLE	1/2Base X Height	1/2Base X Height X Width
CYLENDER	Pi X Radius sqrd.	Pi X Radius sqrd. X Height
SPHERE	4Pi X R sqrd.	4/3Pi X R cubed

NUMERIC PREFIXES		
PREFIX	**ABBREVIATION**	**MULTIPLIER**
yocto	y	10-24
zepto	z	10-21
atto	a	10-18
femto	f	10-15
pico	p	10-12
nano	n	10-9
micro	u	10-6
milli	m	10-3
centi	c	10-2
deci	d	10-1
deka	da	101
hecto	h	102
kilo	k	103
mega	M	106
giga	G	109
tera	T	1012
peta	P	1015
exa	E	1018
zetta	Z	1021
yotta	Y	1024
octillio		1027
nonillio		1030

ROMAN NUMERALS			
ROMAN	ARABIC	ROMAN	ARABIC
I	1	LX	60
II	2	LXX	70
III	3	LXXX	80
IV	4	XC	90
V	5	C	100
VI	6	CC	200
VII	7	CCC	300
VIII	8	CD	400
IX	9	D	500
X	10	DC	600
XI	11	DCC	700
XII	12	DCCC	800
XIII	13	CM	900
XIV	14	IM	999
XV	15	M	1,000
XVI	16	MD	1,500
XVII	17	MV	4,000
XVIII	18	V	5,000
XIX	19	X	10,000
XX	20	L	50,000
XXX	30	C	100,000
XL	40	D	500,000
L	50	M	1,000,000

PERIODIC TABLE OF ELEMENTS

IA																	VIIIA
1 H	IIA											111A	IVA	VA	VIA	VIIA	2 He
3 Li	4 Be											5 B	6 C	7 N	8 O	9 F	10 Ne
11 Na	12 Mg	IIIB	IVB	VB	VIB	VIIB	VIII			IB	IIB	13 Al	14 Si	15 P	16 S	17 Cl	18 Ar
19 K	20 Ca	21 Sc	22 Ti	23 V	24 Cr	25 Mn	26 Fe	27 Co	28 Ni	29 Cu	30 Zn	31 Ga	32 Ge	33 As	34 Se	35 Br	36 Kr
37 Rb	38 Sr	39 Y	40 Zr	41 Nb	42 Mo	43 Tc	44 Ru	45 Rh	46 Pd	47 Ag	48 Cd	49 In	50 Sn	51 Sb	52 Te	53 I	54 Xe
55 Cs	56 Ba	57 La	72 Hf	73 Ta	74 W	75 Re	76 Os	77 Ir	78 Pt	79 Au	80 Hg	81 Ti	82 Pb	83 Bi	84 Po	85 At	86 Rn
87 Fr	88 Ra	89 Ac	104 Rf	105 Db	106 Sg	107 Bh	108 Hs	109 Mt									
				58 Ce	59 Pr	60 Nd	61 Pm	62 Sm	63 Eu	64 Gd	65 Tb	66 Dy	67 Ho	68 Er	69 Tm	70 Yb	71 Lu
				90 Th	91 Pa	92 U	93 Np	94 Pu	95 Am	96 Cm	97 Bk	98 Cf	99 Es	100 Fm	101 Md	102 No	103 Lr

Craig Jay Riley

WIRE FERRULES

AWG	mm2	COLOR
24	0.25	Light Blue
22	0.34	Turquoise
20	0.50	Orange
18	0.75	White
18	1.00	Yellow
16	1.50	Red
14	2.50	Blue
12	4.00	Gray
10	6.00	Black
8	10.00	Ivory
6	16.00	Green
4	25.00	Brown
2	35.00	Beige
1	50.00	Olive

CAT-5 CABLE WIRING

STRAIGHT THROUGH WIRING

PIN#	SIGNAL	PAIR	WIRE	WIRE COLOR	PIN#
1	Tx	2	1	White/Orange	1
2	Rx	2	2	Orange	2
3	Tx	3	1	White/Green	3
4	Rx	1	2	Blue	4
5	Tx	1	1	White/Blue	5
6	Rx	3	2	Green	6
7	Tx	4	1	White/Brown	7
8	Rx	4	2	Brown	8

CROSSOVER WIRING

PIN#	SIGNAL	PAIR	WIRE	WIRE COLOR	PIN#
1	Tx	2	1	White/Orange	3
2	Rx	2	2	Orange	6
3	Rx	3	1	Green	1
4	Rx	1	2	Blue	4
5	Tx	1	1	White/Blue	5
6	Tx	3	2	White/Green	2
7	Tx	4	1	White/Brown	7
8	Rx	4	2	Brown	8

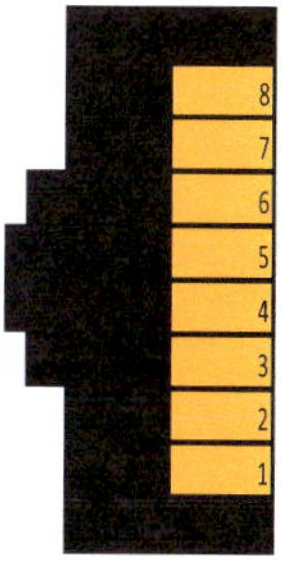

RS-232 PINOUTS		
DB25 PIN	DB8 PIN	DESCRIPTION
8	1	DCD DATA CARRIER DETECT (In)
3	2	RD RECEIVE DATA (In)
2	3	TD TRANSMIT DATA (Out)
20	4	DTR DATA TERMINAL READY (Out)
7	5	SG SIGNAL GROUND
6	6	DSR DATA SET READY (In)
4	7	RTS REQUEST TO SEND (Out)
5	8	CTS CLEAR TO SEND (In)
22	9	RI RING INDICATOR (In)
1		FG FRAME GROUND
9		+ V (In)
10		-V (In)
11		QM (also called SSD- SECONDARY SEND DATA) (I
12		SDCD SECONDARY CARRIER DETECT (In)
13		SCTS SECONDARY CLEAR TO SEND (In)
14		STD SECONDARY TRANSMIT DATA (Out)
15		TC TRANSMITTER CLOCK (In)
16		SRD SECONDARY RECEIVE DATA (In)
17		RC RECIEVER CLOCK (In)
18		UNUSED
19		SRTS SECONDARY REQUEST TO SEND (Out)
21		SQ SIGNAL QUALITY DETECT (In)
23		DATA RATE SELECT (Out)
24		(TC) EXTERNAL TRANSMITTER CLOCK (Out)
25		UNUSED

TAKE UP OF WELDED ELBOWS AND TEES				
	90		45	TEE
	CENTER TO FACE		CENTER TO FACE	CENTER TO END
PIPE SIZE	LONG RADIUS	SHORT RADIUS	LONG RADIUS	
1/2	1 1/2		5/8	1
3/4	1 1/8		7/16	1 1/8
1	1 1/2	1	7/8	1 1/2
1 1/4	1 7/8	1 1/4	1	1 7/8
1 1/2	2 1/4	1 1/2	1 1/8	2 1/4
2	3	2	1 1/4	2 1/2
2 1/2	3 3/4	2 1/2	1 3/4	3
3	4 1/2	3	1 7/8	3 3/8
3 1/2	5 1/4	3 1/2	2 1/4	3 3/4
4	6	4	2 1/2	4 1/8
5	7 1/2	5	3 1/8	4 7/8
6	9	6	3 3/4	5 5/8
8	12	8	5	7
10	15	10	6 1/4	8 1/2
12	18	12	7 1/2	10
14	21	14	8 3/4	11
16	24	16	10	12
18	27	18	11 1/4	13 1/2
20	30	20	12 1/2	15
22	33		13 1/2	16 1/2
24	36		15	17
26	39		16	19 1/2
30	45		18 1/2	22
34	51		21	25
36	54		22 1/4	26
42	63		26	

DIMENSIONS OF 150# RAISED FACE WELDED NECK FLANGES

PIPE SIZE	TAKE UP	No. of HOLES	Dia. Of BOLTS	BOLT CIRCLE	STUD LENGTH	WEIGHT(lbs)
1/2	1 7/8	4	1/2	2 3/8	2 1/2	2
3/4	2 1/16	4	1/2	2 3/4	2 3/4	2
1	2 3/16	4	1/2	3 1/8	2 3/4	2 1/2
1 1/4	2 1/4	4	1/2	3 1/2	3	2 1/2
1 1/2	2 7/16	4	1/2	3 7/8	3	4
2	2 1/2	4	5/8	4 3/4	3 1/2	6
2 1/2	2 3/4	4	5/8	5 1/2	3 3/4	10
3	2 3/4	4	5/8	6	3 3/4	11 1/2
3 1/2	2 13/16	8	5/8	7	3 3/4	12
4	3	8	5/8	7 1/2	3 3/4	15
5	3 1/2	8	3/4	8 1/2	4	19
6	3 1/2	8	3/4	9 1/2	4 1/4	24
8	4	8	3/4	11 3/4	4 1/2	39
10	4	12	7/8	14 1/4	5	52
12	4 1/2	12	7/8	17	5	80
14	5	12	1	18 3/4	5 3/4	102
16	5	16	1	21 1/4	5 3/4	127
18	5 1/2	16	1 1/8	22 3/4	6 1/4	140
20	5 11/16	20	1 1/8	25	6 3/4	170
22	5 7/8	20	1 1/4	27 1/4	7 1/4	224
24	6	20	1 1/4	29 1/2	7 1/2	260
26	5	24	1 1/4	31 3/4	7 3/4	260
30	5 1/8	28	1 1/4	36	8	338
34	5 15/16	32	1 1/2	40 1/2	8 3/4	468
36	6 3/8	32	1 1/2	42 3/4	9	534
42	7 5/8	36	1 1/2	49 1/2	9 1/2	788

DIMENSIONS OF 300# RAISED FACE WELDED NECK FLANGES						
PIPE SIZE	TAKE UP	No. of HOLES	Dia. Of BOLTS	BOLT CIRCLE	STUD LENGTH	WEIGHT(lbs)
1/2	2 1/16	4	1/2	2 5/8	2 3/4	2
3/4	2 1/4	4	5/8	3 1/4	3 1/4	3
1	2 7/16	4	5/8	3 1/2	3 1/4	4
1 1/4	2 9/16	4	5/8	3 7/8	3 1/2	6
1 1/2	2 11/16	4	3/4	4 1/2	3 3/4	8
2	2 3/4	8	5/8	5	3 3/4	9
2 1/2	3	8	3/4	5 7/8	4 1/4	12
3	3 1/8	8	3/4	6 5/8	4 1/2	15
3 1/2	3 3/16	8	3/4	7 1/4	4 1/2	18
4	3 3/8	8	3/4	7 7/8	4 3/4	25
5	3 7/8	8	3/4	9 1/4	5	32
6	3 7/8	12	3/4	10 5/8	5	42
8	4 3/4	12	7/8	13	5 3/4	67
10	4 5/8	16	1	15 1/4	6 3/4	91
12	5 1/8	16	1 1/8	17 3/4	7 1/4	138
14	5 5/8	20	1 1/8	20 1/4	7 1/2	186
16	5 3/4	20	1 1/4	22 1/2	8	246
18	6 1/4	24	1 1/4	24 3/4	8 1/4	305
20	6 3/8	24	1 1/4	27	8 3/4	378
22	6 1/2	24	1 1/2	29 1/4	9 1/2	429
24	6 5/8	24	1 1/2	32	9 3/4	545
26	7 1/4	28	1 5/8	34 1/2	10 3/4	615
30	8 1/4	28	1 3/4	39 1/4	12	858
34	9 1/8	28	1 7/8	43 1/2	13	1110
36	9 1/2	32	2	46	13 3/4	1233
42	10 7/8	36	2	52 3/4	14 3/4	1739

DIMENSIONS OF STANDARD PIPE			
Nominal Size Inches	Outer Diameter Inches	Inner Diameter Inches	Wall Thickness Inches
2	2.375	2.067	0.154
3	3.500	3.068	0.216
4	4.500	4.026	0.237
6	6.625	6.065	0.280
8	8.625	8.071	0.277
10	10.750	10.136	0.307
12	12.750	12.090	0.330
16	16.000	15.250	0.375

DIMENSIONS OF SEAMLESS AND WELDED STEEL PIPE					
		Schedule 40		Schedule 80	
Pipe Size	O.D.	Lbs per FT.	Wall Thickness	Lbs. per Ft.	Wall Thickness
1	1.320	1.7	0.130	2.2	0.18
1 1/2	1.900	2.7	0.140	3.6	0.20
2	2.375	3.7	0.150	5.0	0.22
2 1/2	2.875	5.8	0.200	7.7	0.27
3	3.500	7.6	0.216	10.3	0.30
4	4.500	10.7	0.230	15.0	0.34
6	6.625	19.0	0.280	28.6	0.43
8	8.625	28.5	0.320	43.4	0.50
10	10.750	40.5	0.360	64.3	0.59
12	12.750	53.5	0.400	88.5	0.69
14	14.000	63.5	0.440	106.0	0.75
16	16.000	82.7	0.500	136.4	0.84
18	18.000	104.7	0.560	170.7	0.94
20	20.000	122.9	0.590	208.9	1.03
24	24.000	171.0	0.690	296.3	1.22

STAINLESS STEEL TUBING

TUBE SIZE O.D.	WALL THICKNESS	INNER DIAMETER
1/4	0.035	0.180
	0.065	0.120
	0.083	0.084
3/8	0.035	0.305
	0.065	0.245
	0.083	0.209
1/2	0.035	0.430
	0.065	0.370
	0.083	0.334

POLY TUBING COLOR CODE

SERVICE	COLOR
AIR SUPPLY	RED
TRANSMITTED	ORANGE
CONTROLLED	YELLOW
SEAL	PURPLE
SET	BLACK
ALARM	GREEN
READ OUT	BLUE
ALL OTHERS	NATURAL

The Instrument Tech

He tweaks, he spans, the whole day through.
Problems were ere he goes.
Making the plants run right and true.
Instrumentation is what he knows.

Long hours he toils, both day and night.
Checking and scanning, making sure it's right.
Perfect and true, that's his biz.
Instrument Tech. is what he is.

By Craig Jay Riley

Made in the USA
Las Vegas, NV
07 November 2026

57673305R00052